SPECTRUMS

SPECTRUMS

Our Mind-boggling Universe from Infinitesimal to Infinity

David Blatner

BLOOMSBURY

NEW YORK · LONDON · NEW DELHI · SYDNEY

Images on pages 36, 37, 43, 89, 119, and 147 courtesy of NASA and the NASA Earth Observatory; page 6, snowflake courtesy Kenneth G. Libbrecht, Caltech; page 13, coins by Proskurina Yuliya, Fotolia.com

Published by Bloomsbury USA, New York

All papers used by Bloomsbury USA are natural, recyclable products made from wood grown in well-managed forests. The manufacturing processes conform to the environmental regulations of the country of origin.

LIBRARY OF CONGRESS CATALOGING-IN-PUBLICATION DATA

Blatner, David.
Spectrums: our mind-boggling universe from infinitesimal to infinity / David Blatner.
p. cm.
ISBN: 978-0-8027-1770-2 (hardback)
1. Spectrum analysis. I. Title.
QC451.B53 2012
539.2—dc23
2012010727

1 3 5 7 9 10 8 6 4 2

Designed by Scott Citron
www.scottcitrondesign.com

First U.S. edition published by Walker & Co. in 2012
This paperback edition published in 2014

Paperback ISBN: 978-1-62040-520-8

Printed in the United States by RR Donnelley & Co, Harrisonburg, Virginia

SPECTRUMS

To Gabriel and Daniel,
who help me keep everything in perspective

Contents

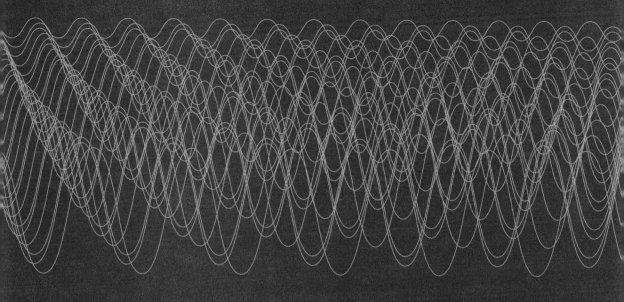

Shades of Anger

Limpid
Quiescent
Tranquil
Phlegmatic
Composed
Mellow
Alert
Anxious
Fretful
Huffy
Grouchy
Cranky
Snappish
Stewed
Peevish
Chafed
Roiled
Mad
Indignant
Boiling
Irate
Enraged
Rabid
Possessed
Wrathful
Apoplectic
Nuclear

INTRODUCTION

**Science and art share a common mandate—
to find surprise in the ordinary by seeing it from
an unexpected point of view.**
—Howard Bloom

MY MOTHER MEANT WELL. "NEVER COMPARE YOURSELF TO others," she taught me. "It'll do nothing but bring you misery." I see her point, but comparing ourselves—whether to other people or to other things—is what we humans do. We compare; we contrast. We sift and segregate, then order, analyzing similarities and discovering differences. That's why we have eyes, ears, taste buds, nerve cells, brains: to compare ourselves with things larger or smaller, faster or slower, hotter or cooler. Through comparisons we create our understanding of the incredibly complex world around us, we realize aesthetic beauty, and we structure our societies.

When you compare two things, you create a duality: this or that, zero or one, black or white. But compare four, or eight, or sixteen things, and you begin to find spectrums—ranges, or continuums *from* this *to* that, a bit darker or lighter, a tad louder or more quiet. The more careful your observation, and the more sensitive your tools of measurement, the better you can gauge—and engage with—your environment.

I attempt, in this book, to provide a sense of scale across six spectrums with which we interact every day: numbers, size, light, sound, heat, and time. Certainly, there are many more spectrums we could explore—density, weight, chemical concentration (which we can sense through smell and taste), and so on—but these six are

"My religion consists of a humble admiration of the illimitable superior spirit who reveals himself in the slight details we are able to perceive with our frail and feeble mind."
—Albert Einstein, physicist

among the best understood in science and, themselves, represent a good spectrum of our everyday experience. True, in the process of discovery, we'll find that we must compare ourselves with others, but rather than misery, I believe we'll find awe, astonishment, and a sense of humility.

Human Scale In his classic novel *The Hitchhiker's Guide to the Galaxy*, Douglas Adams tells the tale of a massive fleet of alien spacecraft that attack Earth, only to discover that they have made a critical error in scale just before landing in a park and being eaten by a small dog. While none of us (I hope) is planning an interplanetary invasion, the lesson remains valid: It's crucial that we constantly evaluate our own perspectives and assumptions as we interact with the world.

Unfortunately, we tend to base our sense of reality on our own human scale and ignore the invisible and often surprisingly nonintuitive worlds beyond. As the biologist Richard Dawkins notes, "Our brains have evolved to help us survive within the orders of magnitude of size and speed which our bodies operate at." We're comfortable within these realms, which Dawkins calls the "middle world . . . the narrow range of reality which we judge to be normal, as opposed to the queerness of the very small, the very large and the very fast." The middle world encompasses distances easily walked, times and durations within our average life span, and temperatures more or less within the range we experience here on Earth, from ice to inferno.

No doubt it's crucial to consider these human scales in the fields of architecture, ergonomic design, retail, and entertainment. But for millennia, we've intuited that there is something more, beyond our senses, both greater and deeper. The idea that we live in a somewhat muted "middle world" inspired both shamanic traditions and many of our greatest myths. The ancient Greeks and Hindus described us as sandwiched by worlds above and below, which are populated by gods and demons. The Christian and Nordic theologies told of realms

I use the plural **spectrums** instead of *spectra* for a reason: While both are correct, the latter has unfortunately gained the supernatural connotation of ghosts and spirits. Although *spectrum*, *spectra*, *spectral*, and *specter* all derive from the Latin word for "vision," we often apply them to phenomena that have nothing to do with what we can see. This is similar to how, over the last century, a number of words with roots in science—such as *dimension*, *evolution*, and even *energy*—have expanded their meanings far beyond their original intention. Today, the word *spectrum* indicates virtually any broad range of characteristics or ideas.

to which we middle-worlders aspire (or fear), just beyond our mortal veil—worlds higher and lower, transcendent and immanent.

And true enough, as we look through microscopes, we find worlds within worlds, realms that explain our everyday rules of physics and biology, but in which, amazingly, those same rules don't always apply. As we peer through telescopes—not just at visible light, but also at the invisible wash of X-rays, gamma rays, and microwaves that we can detect—we discover a universe grander, and weirder, than anything we had imagined.

Science has led us to realize that there is far more outside our human scale than within it, and there is actually very little in this universe that we can feel, touch, see, hear, or possibly even comprehend. The vast range of phenomena around us boggles the mind. It's not an easy task to stretch our imagination to encompass both billions of years and billionths of seconds, or trillions of atoms and trillionths of a meter. And yet, we must explore all these spectrums to gain perspective on our place in the universe.

Spectrums, Everywhere Spectrums—not just the physical spectrums discussed in the chapters that follow but also the very concept of spectrums in general—are useful for far more than just analyzing the world in which we live. Spectrums allow us to communicate our insights and desires, no matter how mundane, with others.

If you're trying to describe a color to a graphic designer, for example, you might explain, "yellow, a bright sunshiny yellow like a pound of butter." In that one phrase you invoke two spectrums: hue and brightness (or, technically, frequency and amplitude). Each spectrum is like a model, or a map, that reflects a dimension of the world around us or within us. So if you hear a specific note played on a piano or violin, you can probably imagine another tone a little higher or lower. Sip your coffee and you viscerally know it would taste better if it were warmer. These daily comparisons may seem minor, but the fact that we can build internal maps like this, often

> "My task is to convince you not to turn away because it appears incomprehensible . . . You see, my physics students don't understand it . . . because the professor doesn't understand it. Nobody does. The thing that is exciting about this is that nature is as strange as can be! The rules of nature are so screwy you can't believe them . . . I don't understand it either! But the fun of it is that it's so mysterious!"
>
> —Richard Feynman, Nobel Prize–winning physicist

extrapolating beyond the ranges of our own personal experience, is an amazing example of what the human mind can do.

Of course, even tiny bacteria show their preferences, or fundamental tendencies, based on simple comparisons between more or less acidic environments. But as life becomes more complex, the more you can name spectrums, or categories, and the more you can name differences in those spectrums, the more you can comprehend, communicate, and thrive. After all, would you rather draw with a box of eight crayons or the jumbo 64-pack?

These spectrums can describe far more than physical measurement; we can use them to describe personal values and aesthetics, too. Any psychologist will tell you that if you're trying to explain how angry you are, it's more helpful to say "I'm about an eight on a scale of one to ten" than to insist on the binary "I'm angry" or "I'm not." The same rule applies, of course, to sports competitions such as Olympic ice dancing, where judges rate technical ability and style against an agreed-upon spectrum—everyone has an intuitive understanding of the difference between a 7.4 and a 7.8; one is just *that much* better.

Granted, not everything fits on a spectrum. In music, you can place rhythm on a spectrum, but not the timbre of individual instruments. Timbre—that extraordinary combination of overtones and harmonics that allows us to tell the difference between an oboe and a violin, even when they play the same note—is a quality, a texture, a shape in and of itself. A circle cannot be more or less circular; a Picasso cannot be more or less Picasso-ish. A sequence is similarly not a spectrum; sure, you can inscribe the life cycle of a butterfly along a line, and you can attempt to identify a caterpillar you've found against that ruler, but there isn't a single parameter, or dimension, you can look at, so there's no "spectrum of butterflyness."

Our richest experiences derive from the collisions of two or more spectrums. Music, for example, is an astonishingly complex

> **"We don't see things as they are, we see them as we are."**
> —Anaïs Nin

interaction among frequency (tone), amplitude (loudness), timing (rhythm), and more. But music itself is not a spectrum.

On the other hand, many ranges that certainly are spectrums are either completely subjective or extremely difficult to measure. Take humor. There's no doubt that one joke is funnier to you than another, but a joke that makes you snort milk out your nose is certain not to be funny to someone else; you might not even find it funny yourself in the presence of your children or a parent. Psychiatrists measure autism along a spectrum, and philosophers measure a spectrum of consciousness, but these are prone to debate (is a dolphin more conscious than a horse?) and require constant reevaluation.

Ultimately, although spectrums are important, they're no more "true" than any map or model, and it's easy to become fooled by looking at the map incorrectly, or through the lens of our own limited understanding. The sun and the moon appear about the same size in the sky, even though one has a diameter 400 times larger than the other. Similarly, it takes just a minuscule concentration of ammonia in the air for you to gasp in shock, while many other compounds hardly register to your nose, no matter how much you sniff. Clearly, when discussing spectrums, we must stand back and look at not just our experience but also how we measure and interpret what we sense.

Infinite Sensitivity We've come a long way in a short time. Just two centuries ago, we didn't realize that electricity had something to do with magnetism, and only a century ago we weren't sure if other galaxies existed beyond our own. But throughout the twentieth century, each decade saw breakthroughs beyond the previously assumed limit of one spectrum or another—radically expanding the age of the universe and the number of galaxies, measuring the size of an atom, smashing the sound barrier, or quantifying the unfathomable energy from an exploding star. Now, in the twenty-first century, we continue to make progress in leaps and bounds,

> "Have you ever noticed that anybody driving slower than you is an idiot, and anyone going faster than you is a maniac?"
> —George Carlin

but rather than pushing the outside limits of a spectrum, the breakthroughs tend to be increased sensitivity, finding exponentially smaller differences in measurements.

But how much is enough? Surely a millionth of a second makes no difference when playing soccer, nor an error of a billion years when discussing how much longer Earth will be habitable. It's obvious, but is worth stating anyway: There are optimal Goldilocks ranges when discussing any spectrum—both in range and sensitivity between values—based on who is talking, and about what.

Watch a professional musician fiddle with an instrument, and you'll be amazed at the small nuances in pitch or tone that she hears and that you simply don't. To you, such subtle distinctions don't matter, but the musician's more sensitive ear enables her to create a richer and more pleasing soundscape.

In fact, almost any expertise involves a deepening sensitivity to one or more spectrums. A typographer looks at a news headline and just feels that it would be more balanced if two letters were a tenth of a millimeter closer to each other. A photographer can see when

▲ Consider a single snowflake out of billions, just 3 mm from tip to tip.

a color print needs 1 percent more magenta. A perfumer knows when a fragrance should shift from citrus to dry woods, or toward greener notes.

As astonishing as this degree of awareness can seem, almost anyone can learn these skills. In fact, researchers have shown that people can even learn how to sniff out a scent trail almost as well as dogs. (Hint: Smelly molecules are heavier than air, so you need to get your nose to the ground to detect them.) The key to developing any of these abilities is simply feeling strong enough about it to try. The same is true in science: The intricate biological mechanisms that take place inside a cell's mitochondrion may be mind-blowing to one biologist and inconsequential to another who is focused on transcontinental ecosystems.

As the English anthropologist Gregory Bateson wrote, "Information is a difference that makes a difference." In order to increase our sensitivity, our abilities, we need first to wake up to the fact that there are differences. This is trickier than it seems; I met a man who told me that after ten years in the printing industry he had only recently noticed that there's a difference between the Times and Helvetica fonts—it had simply not been relevant to his work before. But when he was asked to create a flyer for a customer, what an incredible range of typeface design suddenly began to open to him!

You may even be married to someone who perceives differences that you don't, as she explains her discomfort over how a mutual friend is acting in a social situation, or how fast you're driving. It's tempting to grumble about being "too sensitive," but what if the truth is simply that you lack her awareness of differences in detail on that particular spectrum?

When the Hubble Space Telescope began sending images back to Earth, astronomers were stunned by not only the clarity but also the previously unknown range of the universe. What was a blurry blob in the dark sky suddenly became a crisp set of multiple galaxies, and both our understanding and our expectations grew. We believe that there is even more now—more out there, and more inside,

> **"When you are put into the Vortex you are given just one momentary glimpse of the entire unimaginable infinity of creation, and somewhere in it a tiny little marker, a microscopic dot on a microscopic dot, which says, 'You are here.' . . . In an infinite universe, the one thing sentient life cannot afford to have is a sense of proportion."**
>
> —Douglas Adams, *The Hitchhiker's Guide to the Galaxy*

too, in the realms of the subatomic, the superfast, or even the most mysterious dimension of all: the mind. What if we could develop a telescope in space with a thousand times the sensitivity of Hubble? Or an instrument a trillion times more sensitive than our most advanced microscope? The interfaith campaigner Wayne Teasdale describes God as "infinite sensitivity." Surely, whatever your sense of faith or science, it's a dream to aim for.

Paths to Wonderment Before we launch into our exploration, I have to warn you: Each of these spectrums ranges from interesting to awe-inspiring to truly vertigo-inducing. It's hard to imagine how amazingly small and seemingly insignificant we are in the universe. It's also hard to imagine how amazingly huge and overwhelmingly significant we are in the universe.

Clearly, you need your feet on the ground if you're going to reach for the stars. But where do we find a firm foundation on which to stand? It's a tricky question because—and I know it's not a particularly polite thing for an author to say at the beginning of a book—you really have no idea what's going on. I don't intend any disrespect to you, the reader; it's simply that none of us does. Here's the problem: In our hubris, we humans want to know, we think we *can* know, the whole of the universe. But we can't.

For example, it's obvious that objects have mass, heft, weight, but we don't really know why. Yes, gravity of course, but we don't know why gravity works! It's a mystery. Why does time appear to move forward rather than backward? Is it true that a hundred billion billion times smaller than the proton inside an atom there are just wiggly little strings of energy? We don't know, and it's increasingly clear that we may never know.

We may be like the old story of the blind men and the elephant, where each blind man describes a different aspect of the elephant— the tail, the trunk, the feet—but none of them can understand the whole. As thoroughly as we search the universe for answers, as finely as we tune our instruments, we may be equipped to see only a few parts of the "whole beast."

The result is that as we stretch toward the extremes searching for answers, we find little but more questions, and our most advanced science begins to read like a tract on philosophy or a science-fiction novel. Is there really such a thing as three-dimensional space, as we've all been brought up to believe, or are time and distance a façade, like a hologram that cleverly disguises a far stranger set of realities? It's possible that the Hindus and Buddhists have been right all along, and our universe is all *maya*, or illusion; or that Plato's allegory of the cave in which we see only shadows cast on a wall truly reflects the limits of our senses, and perhaps even our ability to comprehend what we sense.

Nevertheless, just because we can't know it all doesn't mean we can't keep trying and—along the way—enjoy the sense of wonder, awe, delight, and recognition that we are each an integral part of this spectrum of mystery. If we retain a sense of curiosity, and "become like little children" without fear of exploration or shame that we don't yet have the answers, we are bound to learn more, discover more, and slowly develop our abilities over time—exploring the range and depth of our spectrums, opening to their possibilities.

> **"Man is equally incapable of seeing the nothingness from which he emerges and the infinity in which he is engulfed."**
>
> Blaise Pascal, 17th-century mathematician

NUMBERS

None of us really understands what's going on with all these numbers.

—David Stockman, budget director in the Reagan administration

WE ALL LIKE TO THINK WE HAVE SOME SENSE FOR NUMBERS. You may not have enjoyed math in school, but chances are you can peruse your bank statement, squint at a thermometer reading, or count the weeks until Christmas.

We humans are good at these kinds of small numbers. But as keen and insightful as we consider ourselves, as able at digit juggling and mathematical maneuvers as we may be, we are still constrained by our own human limitations: Big numbers (and very tiny ones) are our weakness.

Nevertheless, we are subjected daily to these outliers, especially when exploring the sorts of spectrums discussed in the rest of this book. Even if we can't truly understand the biggest and smallest values, we can at least gain a better intuitive grasp of them.

However, it's no use starting up in the stratosphere, with billions or trillions. We must, instead, start small.

Imagine one: one button, or one person, or one of anything that comes to mind. It's easy to imagine one, or three, or even as many as seven. When imagining a small number of objects, you're likely to see or experience the group as a pattern—perhaps a triangle, a pentagon, or a cross. The technical term is *subitize*: You know the number, you feel it, even without counting.

Beyond six or seven, however, you need to count, or to group small, basic patterns into collections. For example, to conceive "ten,"

you might see two groups of five; to think about "fifty," you might lay out five of those groups of ten.

But unfortunately, it doesn't take a number much larger than that before our human minds get blurry and imprecise. We begin to approximate values, or compare them to a known quantity. For example, asked to consider one thousand, you might picture the seats in a midsized auditorium.

But that's very different from the way you grasped "three." Three is concrete, innate, elemental. Three is an easy pattern, and if there is one thing humans are good at, it's recognizing patterns. Studies show that even human babies on their first day of life are capable of understanding small abstract numbers, correctly matching a pattern of repetitive audio signals to visual stimuli. Australian Aboriginal children who speak Warlpiri (which contains the number words for only *one*, *two*, *few*, and *many*) can similarly distinctly identify groups of five or six—they intuitively understand the patterns even without language for them.

The Meaning Behind the Symbols This method of grouping objects into patterns extends to the sometimes odd symbols and conventions that mathematicians have come up with over the years. For example, multiplication is, at its heart, a way to define a pattern. So 5×3 means *add* three groups of five ($5 + 5 + 5$). That is, multiplication is a way of defining a pattern of adding.

Now, how do you define a pattern of multiplying? The answer is the notion of an exponent—a fancy way of saying, "multiply it *this* many times." For example, 5^3 (which you may see as 5^3) means multiply 5 by itself three times: $5 \times 5 \times 5$.

It sounds complex, but you work with exponents every day without thinking about it. Everyone knows that ten pennies are a dime, ten dimes are a dollar, ten dollars are . . . well, ten dollars. You get the idea. That's exponents—take the whole group and multiply it by the same number. So 10^1 is just 10, 10^2 equals 100, 10^3 is 1,000, and so on. Our basic systems of math and finances are all based on

There are more insects in a single square mile of good, fertile soil than there are human beings on Earth.

Using exponents makes math with huge numbers easy, because you can add and subtract instead of multiply and divide. For example, a million (10^6) times a billion (10^9) equals 10^{15} (because 6 + 9 = 15). Conversely, a billion divided by a million equals a thousand (10^3, because 9 − 6 = 3).

exponents—in this case, exponents of the number 10. These are easy to handle because the exponent simply describes the number of zeros after the 1: Ten (10^1) has a 1 followed by one zero, a hundred (10^2) has two zeros, a thousand (10^3) has three zeros, and so on.

We can do this with pictures, too: Imagine five pennies in a row, then repeat that row five times. That's 5^2, or 5×5, or 25 pennies. Now repeat *that* grid 5 times—for example, you might stack four more pennies on top of each one on the first layer, making a "cube" 5 long, 5 wide, and 5 tall—a total of 5^3 or 125 pennies.

This method of writing numbers is key to understanding what's called scientific notation, such as 4.5×10^9. Once you understand that 10^9 is a billion—a 1 with nine zeros, or 1,000,000,000—then you understand the notation to mean "4.5 billion."

▲ 5^2 pennies (top) versus 5^3 pennies (5 x 5 x 5)

> ## "Learning to compare is learning to count."
> —Edward Kasner and James Newman, *Mathematics and the Imagination*

Even in math there are controversies and disagreements. One such is over the naming of numbers such as 10^9. In North America, we use the "short scale" nomenclature, in which 9 zeros equal a "billion." Residents of many other countries rely on the "long scale," referring to this as a "milliard" or "a thousand million." What Americans call a trillion they then refer to as a billion. Fortunately, after several hundred years of confusion, the long scale has been dwindling in usage worldwide since the 1970s. This is for the best, as otherwise scientific papers on pathologies in duck populations might have to discuss "a milliard mallard maladies." In this book, we'll use only the short scale.

How Big Is Big? The tricky (and powerful) thing about exponential numbers is that very small changes in exponents can reflect really huge changes in the value. For example, the difference between 10^2 (100) and 10^3 (1,000) is 900, but the difference between 10^3 and 10^4 is 9,000! Increasing the exponent by just 1 reflects the difference between the length of the state of California (about 770 miles) and the diameter of Earth (about 7,900 miles). Add 1 more to the exponent (10^5) and you're a third of the way (about 79,000 miles) to the moon.

The difference is even greater when you talk about things like volume. For example, let's say you have a "cube" 100 pennies wide by 100 pennies deep by 100 pennies tall, in other words, 10^2 on each side. That's a total of one *million* pennies, or $10,000. Increase the value of the exponent to just 10^3 pennies per side, and the total number of pennies goes up by 999 million, to a *billion* pennies, or $10 million.

This kind of overwhelming "exponential growth" can lead to startling outcomes. There's an old story of a craftsman who presents a king with an exquisite chessboard and asks, in return, for a single grain of rice on the first square, two grains on the second, four on the third, eight on the fourth, and so on, for all 64 squares. It seems like a reasonable request, so the king quickly agrees.

Unfortunately, the king does not understand the power of exponents. Doubling is a way of saying "2 to the n" or 2^n. So the second square requires 2^1 grains of rice, the third has 2^2 (just 4) grains, and the eighth square at the end of the first row requires only 2^7 grains of rice—a meager 128 grains. But keep going . . . the twenty-first square would need to hold over a million grains, and the forty-first square would hold over a trillion. When you finish the math, you find a total of $2^{64} - 1$ grains of rice. (You have to subtract 1 because you're starting with only one grain, technically 2^0, on the first square.) That's 18,446,744,073,709,551,615 (1.84×10^{19}) grains—enough to fill a box 4 miles long, 4 miles wide, and 6 miles high—taller than Mt. Everest.

Of course, the story goes on to explain that the king, while not wise in numbers, is the perfect politician: He proclaims that in order for the craftsman to receive his payment, he needs to count each and every grain given to him. If the man counted one each second, it would take over half a trillion years—about 42 times the age of our universe—to count the grains!

The story explains beautifully the idea of just how huge numbers can become when working with exponents, and why the phrase "the second half of the chessboard" is sometimes used to describe a situation that has grown far out of control.

Number Numbness Unfortunately, somewhere in the first third of the chessboard the human brain develops what the cognitive scientist Douglas R. Hofstadter calls "number numbness." After all, we can see a thousand objects, so we have a sense—albeit an imprecise

> **Computer scientists often** refer to the speed of a computer in "flops"—floating point operations per second. While most computers today can process megaflops or gigaflops (millions or billions of operations), the fastest computers are just this year breaking the petaflop barrier (quadrillions of operations per second—10^{15}).

one—of what that number means. We can even see ten thousand or a hundred thousand things at the same time—visualize a sold-out football game or political rally where you can see nothing but acres of tiny heads.

The upper boundary of our visual perception, however, is between one and ten million—you could print out a large poster with a million dots on it and stand close enough both to see them all and to resolve each one more or less distinctly. The phrase "I'll believe it when I see it," is applicable: As soon as it's no longer possible for us to see these numbers, we tend to lose any sense of meaning for them. So to many people, any *-illion* word is functionally the same, whether it's *million*, *billion*, *trillion*, or *megazillion* (no, that's not actually a number).

The basic incomprehensibility of large numbers wouldn't be such a problem if it weren't for the fact that modern politics and economics (not to mention science and mathematics) require their comprehension. But how can we grasp a billion when it's a thousand times larger than a million, and a trillion is a million million (10^{12})?

Once again, we do better when we use images to grapple with numbers of this magnitude. So imagine a stack of 100 hundred-dollar bills—just $10,000, small enough to fit easily in a jacket pocket. A hundred of those packets—a pile small enough to fit into a small box or a grocery bag—equals $1 million. Add 99 more of these piles and you've got $100 million sitting just over six feet tall on a shipping pallet.

Now to jump to $1 billion, you need ten of those pallets. Is $1 trillion just a little more? No, you'd need a *thousand* more groups of these 10 pallets to reach $1 trillion.

This unbelievable, mind-stretching number is big. Really big. As one blogger put it, "Your brain can't handle its biggitude." Now double it—match each and every one of the trillion pieces of paper with another—and you get only 2 trillion. You don't add a zero to the number—increasing from 10^{12} to 10^{13}—until you reach 10 trillion. This seems eye-rollingly obvious, but the implications are huge: Each

"Not everything that counts can be counted, and not everything that can be counted counts."

—Albert Einstein

Each cell in the human body contains more atoms than there are stars in the Milky Way galaxy.

$100

$10,000

$1,000,000 ($1 million)

$1,000,000,000 ($1 billion)

$1,000,000,000,000 ($1 trillion)

additional zero means duplicate the previous already-inconceivable group 10 times; each additional "illion" means multiply 1,000 times.

There are myriad examples of how much surprisingly larger these numbers are than each other. For example, take the subject of length: You could stroll a million millimeters in less than an hour, but it takes 10 hours to drive a billion millimeters. A trillion millimeters? That's 25 times around Earth.

Or time: A million seconds is about a week and a half. A billion seconds is 30 years, which is a long time until you notice that Neanderthals were walking around a trillion seconds ago (30,000 years).

Or let's look at dollars again (it always seems to come back to money, doesn't it?). A well-paid professional with $100,000 in the bank can feel confident but may aspire to $1 million. Okay, now create a bar graph where 1 inch = $1 million, and the professional can see that her tenth-of-an-inch bar is paltry against the millionaire's. But now compare the millionaire to Warren Buffett's $65 billion: At this scale, Warren's lot would require a graph more than 1 mile (1.6 km) tall!

Looking at these numbers, you can start to get a sense of how minuscule is a million.

Heading farther into the unfathomable, consider that our own Milky Way galaxy holds about 200 billion (2×10^{11}) stars, and according to observations from the Hubble Space Telescope, we know there are almost 150 billion galaxies in the entire universe. That's an extraordinarily large number, but compare it with something ordinary sitting on your own desk: You can store a trillion bits of information on a common computer hard drive—so if each galaxy were represented by a zero or a one, you could store six universes. And even that number pales compared with the approximately 100 trillion (10^{14}) cells you have in your body.

From time immemorial, humans have been asking which is greater, the biblical "grains of sands of the earth" or the number of stars in the sky. Obviously, neither number is countable, but

Prefix	Abbreviation for Meters	Size	Name
yotta-	Ym	10^{24}	septillions, from the Greek *okto* ("eight," as this is the eighth group of thousands—that is, $1{,}000^8$)
zetta-	Zm	10^{21}	sextillions, from the French *sept* ("seven")
exa-	Em	10^{18}	quintillions, from the Greek *hex* ("six")
peta-	Pm	10^{15}	quadrillions, from the Greek *pente* ("five")
tera-	Tm	10^{12}	trillions, from the Greek *teras* ("monster")
giga-	Gm	10^{9}	billions, from the Greek *giga* ("giant")
mega-	Mm	10^{6}	millions, from the Greek word for "great" (Alexander the Great was *Megas Alexandros*)
kilo-	km	10^{3}	thousands, from the Greek *khiloi* ("thousand"); 1 mile equals 1.6 km
hecto-	hm	10^{2}	hundreds, from the Greek *hekaton* ("hundred")
deca-	dam	10^{1}	tens, from the Greek *deka* ("ten")
deci-	dm	10^{-1}	tenths, from the Latin *decimus* ("tenth")
centi-	cm	10^{-2}	hundredths, from the Latin *centum* ("hundred")
milli-	mm	10^{-3}	thousandths, from the Latin *mille* ("thousand"); 1 inch is about 25½ mm
micro-	µm	10^{-6}	millionths, from the Greek *mikros* ("small"); sometimes called microns
nano-	nm	10^{-9}	billionths, from the Latin *nanus* ("dwarf")
pico-	pm	10^{-12}	trillionths, from the Celtic *beccus* ("beak" or "sharp point")
femto-	fm	10^{-15}	quadrillionths, from the Danish *femten* ("fifteen")
atto-	am	10^{-18}	quintillionths, from the Danish *atten* ("eighteen"); note that an attometer is 1,000 zeptometers and a thousandth of a femtometer
zepto-	zm	10^{-21}	sextillionths, from the Latin *septem* ("seven," as this is the seventh group of thousandths—that is, $1{,}000^{-7}$)
yocto-	ym	10^{-24}	septillionths, from the Greek *octo* ("eight")

University of Hawaii researchers (who should know) estimate Earth has about 7.5 billion billion (7.5×10^{18}) grains of sand. And even though we can see no more than a few thousand stars with the unaided eye, astronomers currently think there are about 16 sextillion stars throughout the known universe (give or take a few)—that's 16×10^{21}. That's an unbelievably large number, but it turns out to be about how many molecules you'd find in just ten drops of water.

Yes, molecules really are that small, and of course atoms are even smaller. In only 12 grams (just under half an ounce) of pure carbon 12 there are 6.02×10^{23} atoms. That seemingly arbitrary number happens to be so central to the study of chemistry that it has its own name: Avogadro's constant. This value defines the size of a "mole"—that is, 1 mole of any substance contains exactly that many molecules. Armed with this knowledge, you can glance at a table of chemical weights and find that a single grain of sugar contains a trillion (10^{12}) molecules of sucrose. That's insignificant compared with a grain of salt, which contains just over a quintillion (1.03×10^{18}) molecules of the very compact sodium chloride.

Chunking The idea that you can take a number like 602,214,179,300,000,000,000,000 and give it a simple name like Avogadro is called chunking, and it's what keeps us sane when dealing with extreme values. We chunk numbers all the time: It's far easier to think in terms of dollars than in hundreds of pennies. Even numbers like "a million" are chunks, but no one wants to think about "two million pennies" when you could say "20,000 dollars"—or, even better, chunk it down to the easier "20 grand." Because we have an intuitive understanding of both "20" and "a grand," we don't have to convert it to smaller units. The chunk is a kind of psychological reality; we know what it is, so we can work with it, mulling the number over and deciding whether the car in the ad is worth that much.

Similarly, we use the term "one hertz" instead of "times per second." Then we add prefixes to make ever-bigger chunks, so

> **You have about 25 trillion** (2.5×10^{13}) red blood cells in your body right now.

> **Three moles of water would** fill about a quarter cup. Three moles of M&M candies would fill all the oceans of the world.

kilohertz (kHz) means thousands of times each second, and megahertz (MHz) means millions. (In the example above, it would make far more sense to say "the car costs 2 megacents," but people would look at you funny.)

Astronomers chunk the average distance between Earth and the sun, calling it 1 astronomical unit (1 AU)—easier than writing 150 million kilometers—and they chunk 63,000 AU into a single light-year. You may have no intuitive sense of what a light-year is (the distance light travels through space in one Earth year), but you can certainly understand that an astrophysicist discussing the distance from here to Vega is far happier scribbling 25 ly rather than 147,962,000,000,000 miles or even 2.37×10^{17} meters. Of course, it seems that no matter how large the chunk, the numbers still get out of hand, growing cosmically—one might even say comically—huge. The farthest object ever detected—a gamma ray blast from an exploding star—sits at such a length: some 13,140,000,000 light-years (1.2×10^{26} meters) away.

Beyond the Possible In 1938, the mathematician Edward Kasner asked his nine-year-old nephew Milton to name a number so huge, so out of this world, that it would boggle the mind. The reply was "a googol,"* which he precociously defined as 1 followed by a hundred zeros (10^{100}). Urged toward greater heights, Milton then followed up with the googolplex, which he originally determined would be contrived by writing zeros until your hand got tired but was later standardized as a 1 followed by a googol zeros.

> **Googol + 267 is the first** prime number over a googol.

These numbers are bigger than anything we've encountered so far. In fact, not only can we not imagine a googol of anything, there isn't even a googol of anything to imagine! A googol is bigger than the number of molecules in every substance on our planet; it's more than all the hydrogen atoms in our sun. Astonishingly, when you

*Note that this number is spelled differently from the trademarked name of a certain Internet search company.

count up *every atom in the known universe*, it still comes to only about 10^{81} particles, a quintillion times less than a googol.

So if numbers like a googol (not to mention the even more insanely enormous googolplex) are beyond any correspondence with physical matter, why even bother with them? Because mathematics demands it. While most students' experience of mathematics stops not far beyond arithmetic, professional mathematicians drive farther and delve far deeper, past the shallow solving of equations, in an attempt to understand the underlying nature of numbers and, indeed, the universe itself. To describe the universe—or one of the many potential multiverses—you must go beyond its boundaries, just as the paper you draw on must be larger than the picture you're drawing.

You can't really avoid "very large numbers" (as they're quaintly described in textbooks) when doing higher math or studying code breaking or cosmology. For example, we earlier looked at the meaning of the simple number 4^4, which equals 256. But what about 4^{4^4}? That seems innocuous enough at first, but the answer is actually more than 154 digits long (it's 1.34×10^{154})! Mathematicians call this tetration—as in "4 tetrated to 3"—or sometimes call it superpower, superdegree, or a term nine-year-old Milton would be proud of: hyper4.

Hyper4 numbers take mathematics to a whole different playing field, one beyond the simple number crunching of electronic computers. Sure, a computer can brute-force analyze all the possible moves in a game of chess (the total number of moves in an entire game is probably on the order of 10^{50}), but in the ancient game of Go—played with simple black and white stones on a 19-by-19 grid—there are more than 10^{150} possible positions.

The patterns of logic and number spiral out ever higher. In 1933, the South African mathematician Stanley Skewes was studying how prime numbers* are distributed across the spectrum of numbers

In Darren Aronofsky's movie π, the protagonist, Max, tells a group of Jewish kabbalists that he knows they have written down every possible 216-digit number. Of course, as a mathematician, Max must know this is impossible. Even a million supercomputers working steadily from the big bang until today couldn't achieve that goal.

*A prime number is any number greater than 1 that is evenly divisible by only 1 and itself.

when he published 1.397×10^{316} in his research—a number so large that it was given a name (Skewes' number) and dubbed by the famous mathematician G. H. Hardy as "the largest number which has ever served any definite purpose in mathematics." But the record was not to last, and Skewes' number looks positively quaint compared with today's most advanced math functions, which regularly include numbers such as $10^{10^{600}}$. Very large indeed.

Going Negative It appears to be a universal adage that "as above so below," and this is seen nowhere as clearly as in the world of the number. The inverse of the number 2 is ½—one half of 1. The inverse of 3 is ⅓, smaller than a half. As you raise the number to 4, 5, 10, 100, and greater, its inverse decreases (¹⁄₁₀, ¹⁄₁₀₀, and so on), approaching, yet never reaching, zero. What's smaller than 1/googol? Of course the answer is 1/googolplex!

By the way, you can describe extremely small numbers using much the same notation as very large ones. Where 1×10^3 means "move the decimal point 3 places to the right" (1,000), the notation 1×10^{-3} means move the decimal point 3 places to the left (0.001, or one thousandth, ¹⁄₁,₀₀₀). A millionth is 10^{-6}, a billionth is 10^{-9}, and so on.

But at some point you're bound to hit zero . . . and then what? Just as you cannot count a googol objects, you can't count fewer than zero. The ancient Greeks, who even twenty-five hundred years ago could do far more precise math than you might expect, had a serious weakness: They rejected any numbers that wouldn't fit into their geometries. Can't draw a picture of a number less than zero? Then, in their world, it didn't exist. Granted, it's not an entirely unreasonable assumption: Ask a six-year-old to solve "2 take away 2," and she can tell you; ask her "2 take away 3" and you'll see her little eyebrows screw up in an innocent imitation of Mr. Spock: That does not compute!

However, the Chinese and the Hindus of the pre-Christian era didn't need to represent numbers with pictures or countable objects,

"**Every time you drink a glass of water, you are probably imbibing at least one atom that passed through the bladder of Aristotle. A tantalisingly surprising result, but it follows [from the] observation that there are many more molecules in a glass of water than there are glasses of water in the sea.**"

—Richard Dawkins

and both came up with a fairly radical idea at the time: the negative number. You can't count it, but you know it's there because it makes sense that it should be. As the great mathematician Carl Friedrich Gauss wrote, "Just as in general arithmetic no one would hesitate to admit fractions, although there are so many countable things where a fraction has no meaning, so we would not deny to negative numbers the rights accorded to positives."

So once again we find pairs of numbers: 15 is matched with –15, 10^{261} (sexoctogintillion) is mirrored by -10^{261}, and so on.

And so on? If you stop and think about it, the phrase "and so on" is just as radical a concept as negative numbers: It connotes forever, eternally, the infinite. Here, once again, we are asked to extrapolate beyond the comfortable countable universe. It is a basic premise of our mathematics that there is such a thing as the infinitely large and the infinitely small. But dealing with infinity—known by many terms, including "aleph null" and "the set of N"—is a slippery business not to be undertaken by the faint of heart.

Remember, infinity is not a destination but rather an idea. There is no point where the very large numbers begin to merge subtly into infinity. The astronomer Carl Sagan wrote, "A googolplex is precisely as far from infinity as is the number 1 . . . No matter what number you have in mind, infinity is larger."

Even doing math with infinity takes on a bizarre fun-house-mirror quality. Infinity plus 1 is infinity. Infinity plus infinity equals infinity. While it appears that the list of odd numbers would be half the size of the list of all the numbers, it is not so—rather, both are (you guessed it) infinite. As Philip Davis and Reuben Hersh wrote in *The Mathematical Experience*:

> The set of N is an inexhaustible jar, a miraculous jar recalling the miracle of the loaves and the fishes in Matthew 15:34.
> This miraculous jar with all its magical properties, properties which appear to go against all the experiences of our finite lives, is an absolutely basic object in mathematics, and thought to be well within

the grasp of children in the elementary schools. Mathematics asks us to believe in this miraculous jar and we shan't get far if we don't . . .

The infinite is that which is without end. It is the eternal, the immortal, the self-renewable, the *apeiron* of the Greeks, the *ein-sof* of the Kabbalah.

You could say that plus and minus infinity define the endpoints of the spectrum of numbers, except that, by their very definition, they cannot be endpoints at all.

Thinking Outside the Box With a number line stretching out toward negative and positive infinity, like a railway traversing eternity, we should be able to pinpoint the answer to any mathematical problem we encounter, right? Amazingly, you don't have to search long before you find an equation for which there is no train station on that line, no place at which you can definitively say, "This is the solution."

Instead, we have to deal with numbers like the irrational—those that can be written as an infinitely long decimal sequence but cannot be described simply by dividing two integers. For example, find a number that you can multiply by itself to result in 2—the square root of 2 (or $\sqrt{2}$). We can estimate it by dividing 90 by 63, or we can write it as 1.4142135 . . . But that "dot dot dot" at the end means we can never isolate its value exactly—the digits simply rattle off, without pattern, forever.

Then there are the transcendental numbers. Originally named when mathematicians thought these values were exceedingly rare, we now know they are as common as dust scattered throughout mathematics. A transcendental number is not only irrational but also nonalgebraic. That is, it cannot be described by a simple, finite algebraic equation. Many irrational numbers are also transcendental, such as the famous constants π and e.

However, both irrational and transcendental numbers do live somewhere on the number line, even if we can't put our finger on them precisely. There's another class of number so weird that we have to step off the tracks entirely to grasp it. Let's look at a simple

$\boldsymbol{\pi}$ =

3.14159265358979323846264338327950288419716939937510582097494459230781640628620899862803
4825342117067982148086513282306647093844609550582231725359408128481117450284102701938521
10555964462294895493038196442881097566593344612847564823378678316527120190914564856692346
03486104543266482133936072602491412737245870066063155881748815209209628292540917153643678925903600113
305305488204665213841469519415116094330572703657595919530921861173819326117931051185480744623799627495
6735188575272489122793818301194912983367336244065664308602139494639522473719070217986094370277053921717
6293176752384674818467669045132000568127145263560827785771342757789609173637178721468440901224953430146
54958537105079227968925892354201995611212902196086403441815981362977477130996051870721134999999983729
78049951059731732816096318595024459455346908302642522308253344685035261931188171010003137838752886587533
20838142061717766914730359825349042875546873115956286388235378759375195778185778053217122680661300192
78766111959092164201989380952572010654858632788659361533818279682303019520353018529689957736225994138912
4972177528347913151557485724245150695950829533116861727855889075098381754637464939319255060040092770167113900984882401285836160356370766010471018194295559619894676783744944825537977472684710404753464620
8046684259069491293313677028989152104752162056966024058038150193511253382430035587640247496473263914199272604269922796782354781636009341721641219924586315030286182974555706749838505494588586926995690927217
079750930295532116534498720275596023648066549911988183479775366369807426542527862551818417574672890977
7727938000816470600161452491921732172147723501414419735685481613611573525521334757418494684385233239073
9414333454770246186625189835694855620992192221842725502542568876717904946016534668049886272327917860857
843838279679766814541009538837863609506800642251252051173929848960841284886269456042419652850222106611863067442786220391949450471237137869609563643719172874677646575739624138908658326459958133904780279...

$\boldsymbol{e}$ =

2.71828182845904523536028747135266249775724709369995957496696762772407663035354759457138217852516642742746639193200305992181741359662904357290033429526059563073813232887294349076323382988075319525101901157383418793070215408914993488416750924476146066808226480016847741185374234544243710753907774499206965517027618386062613313845830007520044933826560297606737113200709328709127443747047230696977209310141692836819025515108654746377211125238978442505695369677078544996996794686445490598793163688923009879312773617821542499922957635148220826989519366803318252886939384946464651058209392398294887933203625094431170129681970684161403970198376793206832823764648042953118023328782509819455815301756717361332069811250996181881593041690351598885193458072738667385894228792284998920868058257492796104841984443634632449684875602336624827041978623209002160990235304369941849146314093431781436405462531520961836908887070167683964243781405927145635490613031072085103837505101157477041718986106873969655212671546889570350350354021234078498193343210681701210056278802351930332247450158539047304199577770935036604169973297250886876966403555707162268447162560798826517871341951246652010305921236677194325278675398558944896970969640975459185695638023637016211204774272283648961342251644507818244235294863637214174023889344124796357437026375529444833799801612549227850925772562092622642832627793338656648162775164019105900491644998289315056660472580277863186415519565324425869829469593080191529872117255634754639644791014590409058629849679128740687050489585586717479854667757573205681288459205413340539220001137863009455606881674001698420558040336379537645203040243225661352783695117788386387443966253224985065499588623428189970773327617178392803494650143455889707194258639877275471096295374152111513683506275260232648478700392076431005958411661205452970302364725492966693811513732275364509888903136020572481765851180630364428123149655070475102544650117272115551948668508003685322818315219600037356252794495158284188829478761085263981...

algebraic equation: $x^2 - 1 = 0$. To solve this, we add 1 to both sides and find $x^2 = 1$. In other words: What number multiplied by itself equals 1? The answer is obviously 1. (Technically, the answer could also be –1, because a negative value multiplied by itself always results in a positive.)

Okay, so now let's make a tiny change to the equation, changing the minus to a plus: $x^2 + 1 = 0$. What number when multiplied by itself equals –1? Up go the Spock eyebrows, and clank goes the brain. You could take the easy way out, like the Greeks on negative numbers, and say, "That does not exist." Or instead, you could look into the fog, step off the number line, and use your imagination. As the great mathematician Leonhard Euler wrote, the answers to these sorts of questions "are neither nothing, nor greater than nothing, nor less than nothing; which necessarily constitutes them imaginary."

He wasn't saying they didn't exist; he was literally naming them: imaginary numbers, typically described by the letter i, those living on an alternate spectrum of numbers than the one we're used to. While imaginary numbers and their "complex number" friends (such as "2 + 3i") don't show up on household electronic calculators, they are a standard—indeed, an essential—element in the mathematician's toolbox. Without imaginary and complex numbers, scientists could not figure rocket trajectories or work out quantum dynamics. They exist because they should exist, they must exist in a logical system of math—just like negative numbers exist—not because we can see them or count them. As the brilliant seventeenth-century inventor of calculus, Gottfried Leibniz, wrote, "Imaginary numbers are a fine and wonderful refuge of the Holy Spirit, a sort of amphibian between being and not being."

From a foundation of patterns we have built cathedrals of numbers, from the highest spires to the darkest catacombs. The numbers shimmer in our architecture, expressing both the countable and the ineffable, the real and the imaginary. Imagine one. Then imagine it all.

SIZE

Distance lends enchantment to the view.

—Mark Twain

WHAT IF A SINGLE E. COLI BACTERIUM—ONE OF THE MICROSCOPIC creatures that are currently massed in your digestive tract—suddenly, improbably gained self-awareness and intelligence? It would not, could not, comprehend its place in its world—a human body tens of millions of times larger than itself and yet a world of which it is intimately a part. How then are we to understand our place in a universe that is absurdly large and complex in comparison with our nearly insignificant size?

On the other hand, as we consider that tiny bacterium, we realize that from our human perspective we ourselves are incredibly huge, each of us truly containing multitudes. For there are more than 10 trillion cells in the human body, and ten times that many bacteria living inside it (yes, we are more "they" than "us," at least in number). If each cell were a star, your body would contain hundreds of galaxies.

And upon even closer exploration, we find inside each cell billions and billions of atoms, bound together into molecules of water, DNA, and other structures. Your body contains more atoms than there are stars in the universe.

So perhaps our rank depends entirely on our perspective—we are either teeny or enormous, irrelevant or gods, based on how we look at it.

Of course, part of the trouble with gaining perspective is that from our viewpoint we see only a tiny slice of the whole. Just as our

> **"You shall not pervert justice in measurement of length, weight, or quantity. You shall have true scales, true weights, true measurements of dry and liquid."**
> —Leviticus 19:35–37

eyes can see only a portion of electromagnetic radiation (visible light), or hear a segment of a sound spectrum that extends beyond our reach, we experience size and distance only at our human scale. It is only through our scientific instruments that we start to recognize the worlds within and the heavens around us.

Common Measurements To comprehend size—and, more important here, to discuss it—we must take a detour to explore standards of measurement. The standards have changed over time, of course, in both name and precision. The ancient Hebrews and Egyptians used the cubit (the distance from fingertip to elbow), divided into seven palms of four digits each. Unfortunately, there were a number of different cubits: The Egyptian cubit was around 52 centimeters (20.5 in); the Hebrew cubit was shorter, only about 45 cm (17.7 in).

The Romans, interested in larger scales, developed the *mille passuum*, from which our concept of a "mile" originated. Their "thousand paces" was about 5,000 feet (1,500 m) long, which seems to indicate either really that Romans were really tall or that one pace equaled two steps.

Unfortunately, both of these examples reinforce the common misconception that the ancients were sloppy or ignorant with their math and physics. Not so. Take the case of Eratosthenes of Alexandria, Egypt, who in 240 BCE not only proved that the world was round but also calculated Earth's circumference using stadia (the common length of sports stadiums at the time). His result was within 2 percent (a couple hundred miles) of what we know today to be correct.

The stadium may seem like an odd value, but measurements have always been somewhat arbitrary and usually based on a very human scale. The 12-inch foot appears to have originated by the length of some long-forgotten royal foot (though that, too, likely stretched the royal truth). The post-Roman mile was defined as eight furlongs, each of which described a conveniently sized "furrow long" when plowing.

The only countries that haven't standardized on the metric system are the United States, Myanmar, and Liberia.

There are at least 18 different measurements called the "barrel" (including for beer, oil, cranberries, cornmeal, cement, and brandy).

The acre, from the Latin *ager* ("field"), reflected the land that could be plowed by a yoke of oxen in one day.

And, of course, for most of the first two millennia C.E., every town, province, and industry seemed to boast its own measurement system, often with similar names but very different values. Even today we have to be careful when comparing gallons (the British "imperial" gallon is a fifth larger than the American gallon); pounds (a pound may equal the 16 "avoirdupois" ounces that most Americans use or the 12 "troy" ounces used when weighing precious metals); or miles (the nautical mile is 15 percent longer than one on land).

So in the eighteenth century, France, in the heat of the revolution, set out to standardize a metric that everyone could use: The meter (or metre, if you're feeling continental) would be land based, not human based, calculated at one 10-millionth the distance from pole to equator—or, one 40-millionth of Earth's circumference. Unfortunately, as Earth is neither smooth nor truly spherical, even this ideal fell to an arbitrary human compromise. The meter ended up defined as the distance between two lines someone etched on a bar of platinum-iridium alloy—just under 40 inches (3.25 ft).

Desperate to find a true meter in nature, scientists at the 1960 General Conference on Weights and Measures decided to base the meter on the wavelength of a particular orange-red spectrum-emission line of the krypton 86 atom. Still too arbitrary? In 1983, the meter was finally redefined as the "length of the path travelled by light in vacuum during a time interval of 1/299,792,458 of a second." A more self-justified argument for a value would be hard to find, but at least it is a firm standard one can hold on to in these changing times.

The greatest benefit of the meter, of course, is not its definition but rather our ability to use the unit at any scale. The simple, though sometimes obscure, prefixes you can attach to the word make it easy to describe exponential differences in size. For example, add *kilo-* to describe a thousand (10^3) meters. Change it to *mega-* to describe something a thousand times longer (10^6, or a million meters). Or

> "Metric is definitely communist. One monetary system, one language, one weight and measurement system, one world—all communist! We know the West was won by the inch, foot, yard, and mile."
>
> —Dean Krakel, director of the National Cowboy Hall of Fame

Few Americans know that America has been debating the adoption of metric measurements for more than 200 years. To argue the use of meters vs. miles is as American as apple pie and George Washington, who brought the discussion to the public's attention in his first inaugural address. Even fewer Americans know that even the foot and yard are legally based on the value of the meter. One yard equals exactly 0.9144 meter.

switch it to *micro-* to describe 10^{-6} meters—that is, millionths of a meter.

Getting Big, on Earth We have an innate comprehension of the world that we can see and easily measure. That is, we "get" things that are larger than a mote of dust and smaller than, say, our neighborhood. These boundaries define what's known as "human scale," easily measured from thousandths to hundreds of meters.

You're probably shorter than two meters in height, though some people are taller—the American Robert Wadlow (who died in 1940 at the age of twenty-two) broke the record at 2.72 meters (8 ft 11 in). The tallest animal, the giraffe, can grow as high as 5.5 m (18 ft). The tallest living tree is about 115 m (378 ft). The longest baseball throw on record was about 136 m (446 ft), which also happens to be about the height of the Great Pyramid at Giza—the tallest structure in the world for almost four thousand years.

There are no human-made objects that reach as high as 1 kilometer (1,000 m); the Empire State Building is only 381 meters (1,250 ft) tall; and the tallest structure in the world at the time of this writing, the Burj Khalifa skyscraper in Dubai, is only 828 m (2,717 ft). Seeing it rise majestically into the haze evokes the dream of the Tower of Babel, which was to reach to the heavens themselves.

Of course, a kilometer is impressively high, but the same measurement doesn't pack the same punch when describing a length—the Golden Gate Bridge spans 1.28 km from one tower to the next, and the biggest bridges are far longer.

And yet, these human-scale measurements are all easy for us to comprehend, largely because we have personal experiences with things this size. The phrase "seeing is believing" has a kernel of truth, as we tend to deeply understand—to *grok*, using a word coined by the science-fiction author Robert Heinlein—that which we can see. We can see a tree, walk around it, or even climb it, and thereby get an internal understanding of its size in relation to our own.

The world record for the shortest human is held by the 72-year-old Nepalese Chandra Bahadur Dangi at 21.5 inches (54.6 cm) tall.

The Washington Monument in D.C. was built in 1884 to be 555 ft 5 in. (170 m) tall.

However, as we look at objects longer than a kilometer or two, we tend to lose track of scale. We may be able to see it, to believe it, but we don't know it in the same way we can know that tree, because we can no longer correlate it to our own human scale.

That's not to say we don't know it in a different way. After all, it's possible to know a forest better than you know an individual tree, just as you know the tree better than its trillions of constituent cells. What we understand—what we see—is always just a portion of the whole picture, as each object is both itself a complex system and part of an even more complex system. We know the shape of our continent from satellite photography, but that does not imply we comprehend how truly large that landmass is. Our understanding, too, exists on a spectrum.

If we're standing on level ground, Earth's natural curvature prevents us from seeing farther than the horizon, about 5 kilometers (3 mi) away, though from a small mountaintop you can see 190 km (120 mi) or more. Mount Everest reaches only about 9 km above sea level. Few people realize that the Hawaiian island mountain of Mauna Kea is actually almost a mile taller than Everest, but the base of the mountain rests 19,684 feet (6 km) below the waves. The deepest ocean trenches extend far lower, at about 10 or 11 km below sea level.

Most weather occurs within the troposphere, up to 11 km (36,000 ft) from the surface of Earth, though the highest clouds may extend about 24 km (80,000 ft) into the stratosphere. We have a sense that our atmosphere is a huge cushion around us, protecting us from the harsh vacuum and radiation of space. However, think of it this way: If Earth were shrunk to the size of a wet tennis ball, our atmosphere would be no thicker than the water clinging to the surface.

Returning to sea level, it's slightly easier to wrap our heads around these sizes when considering geographic landmarks. Manhattan island is about 22 km (22,000 m, or 72,000 ft) long. You can measure the distance from San Francisco to Honolulu as 3,860 km, but at this

The largest living thing on Earth: the 2,400-year-old, 2,200-acre (8.9 km²) giant fungus living underground in Oregon.

length it starts to become easier to measure in the millions of meters, so we write 3.86 Mm. Flying from Chicago to Tokyo covers 10.16 Mm, almost the diameter of Earth itself (12.75 Mm). The circumference of that big blue ball we call home is a convenient-to-remember 40 Mm (40 million meters . . . well, technically 40,075,160 m around the equator, but who's counting?).

It's impossible for us to simply grasp Earth's size as an absolute; instead, we must rely on comparing its size with objects around us. It would take 14,615 Golden Gate bridges to circumscribe the globe. And you could stack 15,400 Burj Khalifas to span from Beijing through the center of Earth to its antipode, Buenos Aires.

Of course, if we have trouble grokking our own planet, what's to become of us as we leave the comfort of our terrestrial home and begin to explore the size of the cosmos?

The Great Beyond It seems as though space—that ultimate frontier—is unimaginably far away. But jetliners traverse the sky only about 10 kilometers high (about 6 mi), and you need travel only 100 km (62 mi) up before you are no longer considered an aeronaut and are now an astronaut. In fact, the distance from the surface of Earth to

> **The diameter of the Earth** from pole to pole is is just over 500,500,000 inches.

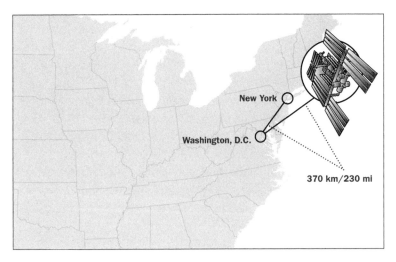

New York

Washington, D.C.

370 km/230 mi

the International Space Station is shorter than the train ride from New York to Washington, D.C., about 370 km (230 mi).

Of course, the hardest part about making "one small step for man" isn't the distance it takes to get into space but rather the force it takes to break free from the bonds of gravity. And ironically, gravity may be the weakest force in the universe! After all, in a battle between the entire mass of Earth and a tiny magnet, the tiny magnet easily wins, picking up a paper clip with ease. Electromagnetism is billions of times stronger than gravity. Nevertheless, it requires tremendous force and speed for a huge rocket to escape the gravitational pull of the planet to reach 250 km (160 mi) from the surface, where it can begin to orbit, or continue on past Earth's exosphere and toward the moon.

It is at this point in our journey outward, however, that the sizes begin to get mind-numbingly large. The exact distance between any two celestial bodies varies, of course, as orbits tend to be elliptical rather than spherical. But the moon is generally around 378 Mm from Earth—that's 378,000 km, about a quarter million miles, or ten times the circumference of our own planet.

The closest planet to us, and the brightest object in the night sky after the moon, is Venus, about 38 billion meters (38 Gm) at its closest—that's more than 100 times the distance between the moon and us. The sun is about 1.4 Gm in diameter, and is, on average, 150 Gm away.

Think about these numbers this way: If Earth were the size of the period at the end of this sentence, the moon would be about 15 millimeters away (about the thickness of your finger), and the sun would be about the size of a child's fist and 6 meters (20 ft) away. In fact, the sun is so large that if Earth were placed at the center of the sun (ignoring the fact that we'd all vaporize), our moon's orbit would reach a bit more than halfway to the solar surface.

When you're discussing planetary distances, it's awkward to count in meters—even gigameters. Instead, a more handy measure is the average distance from Earth to the sun; scientists call this value 1

Voyager 1 set out in 1977, and is currently traveling at about 3.6 AU (538 Gm) per year, or 61,400 km/h. At this rate, it won't reach the Oort cloud for more than 1,000 years, and it will take more than 73,000 years before reaching the nearest star.

▲ Earth as a tiny dot of light, seen through the rings of Saturn from the Cassini spacecraft in 2006.

> "Look again at that dot. That's here. That's home. That's us. On it everyone you love, everyone you know, everyone you ever heard of, every human being who ever was, lived out their lives . . . every king and peasant, every young couple in love, every mother and father, hopeful child, inventor and explorer, every teacher of morals, every corrupt politician, every 'superstar,' every 'supreme leader,' every saint and sinner in the history of our species lived there—on a mote of dust suspended in a sunbeam."
>
> —Carl Sagan

astronomical unit (1 AU). So Jupiter, for example, orbits about 5 AU from the sun.

Jupiter, by the way, is an astonishingly large gaseous planet, with more than twice the mass of all the other planets put together. It's so big that its easily identifiable red dot—thought to be a huge storm that has raged for almost two centuries—is larger than the diameter of Earth. Here's a helpful comparison: If the sun's diameter were equal to the height of a man, Jupiter would be about the size of his head and Earth would be slightly bigger than the iris of his eye.

Heading farther out into our solar system, we find the planets Saturn at just under 10 AU, Uranus at almost 20 AU, and then Neptune at about 30 AU (around 4.5 trillion meters, or terameters) from the sun. Beyond Neptune, stretched across a massive 25 AU of space, is an enormous ring of more than 100,000 rocky icy objects collectively called the Kuiper belt. These chunks of space debris orbiting far from the sun include Pluto and several other frozen dwarf planets, such as Haumea and Makemake. (All large Kuiper belt objects besides Pluto are named after creator deities; Haumea is the

Hawaiian goddess of fertility, and Makemake is the fertility goddess of the Rapanui people of Easter Island.)

The farthest known object in our solar system is Sedna, a rock about two thirds the size of Pluto, which has an extremely elongated orbit. At its closest (the perihelion of its orbit), Sedna is about 76 AU away; after fifty-five hundred years, when Sedna reaches the other end of its orbit (its aphelion), it reaches 937 AU, more than 140 trillion meters into space from the sun. No wonder it was named after an Inuit goddess who is said to live in the darkest, coldest part of the Atlantic Ocean.

Most of us were taught that Pluto marks the outer edge of the solar system, but there is likely far more out there, endlessly orbiting the sun, bound by the same gravitational force that holds our planet in place. Unfortunately, we don't know what else is out beyond the Kuiper belt—it's simply too dark for us to see. Astronomers believe that there is likely nothing at all for an expanse of several thousand AU, spare an occasional comet, followed by a huge swarm of trillions of comets that surround our solar system between 5,000 and 100,000 AU away. This is the theoretical Oort cloud, the outer edge of which is considered the boundary of our solar system.

Light It's all very well and good to bandy about distances like 100,000 AU, but what do they really mean? Once again, the numbers are just too big. So imagine that the solar system (out past the Oort cloud, to the boundary between the solar wind and interstellar space) is the size of a typical elementary school classroom. The sun—by far the largest object in the room—would float in the middle of the classroom, smaller than a grain of salt. Earth, about the size of a microscopic bacterium, would be orbiting about 10 centimeters away.

Or let's flip it around: If Earth were the size of a grain of salt, our solar system (only out to Neptune!) would be 352 meters wide—that's a grain of salt sitting inside about three and a half football fields of space. If you include the whole solar system (out to the Oort cloud), it's more than 2,000 times more space: a grain of salt in a region

▲ The "Blue Marble" photo of Earth was taken by Apollo 17 in 1972, 45 Mm (28,000 mi) from the surface.

If our solar system (out to the Oort cloud) were the size of a grain of salt, the Milky Way galaxy would be about the length of a football field. If the Milky Way galaxy were the size of a grain of salt, the visible universe would be about as large as the 110-story Sears Tower in Chicago.

If the universe were shrunk so that Earth were the size of a period on this page, our nearest star would be 1,500 km (930 mi) away. The center of the Milky Way would be about 8 Gm or 5 million miles away.

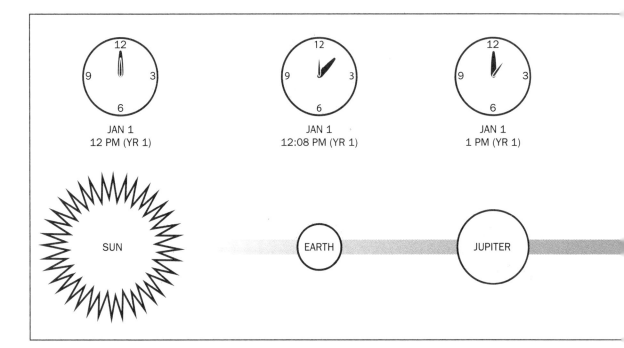

JAN 1
12 PM (YR 1)

JAN 1
12:08 PM (YR 1)

JAN 1
1 PM (YR 1)

SUN

EARTH

JUPITER

about 450 miles wide. (That's like flying from San Francisco to Seattle—a two-hour flight—and encountering virtually nothing but a few specks along the way.)

Of course, no one wants to deal with numbers like 100,000 AU or trillions of meters. There are too many zeros for comfort! Instead, astronomers tend to simplify these enormous values by using parsecs. A parsec is defined in a way that only a mathematician could love, but it's based on the tiny discrepancy we can see when we view distant stars from different positions along Earth's orbit. You're already familiar with this effect, called parallax: Hold your finger up at arm's length and close one eye, then open it and close the other, and you'll see the finger appear to change position against the background scene. Astronomers use parallax to measure distances in parsecs.

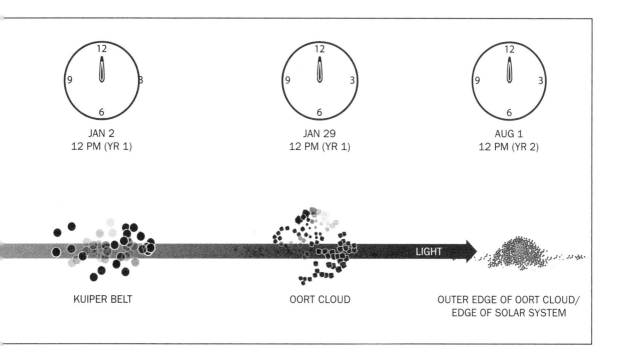

JAN 2
12 PM (YR 1)

JAN 29
12 PM (YR 1)

AUG 1
12 PM (YR 2)

LIGHT

KUIPER BELT

OORT CLOUD

OUTER EDGE OF OORT CLOUD/
EDGE OF SOLAR SYSTEM

The rest of us rely on a much simpler measurement: the light-year. Light always travels at the same speed; in a vacuum, that's almost exactly 299,792,458 meters a second—though it's easier to round up to 300 Mm/sec (which you can write 3×10^8 m/sec). Imagine traveling seven and a half times around the equator in a single light-second.

You can visualize a package of light leaving the surface of the sun at the stroke of 12:00 a.m., January 1. At just after 12:08, it would reach Earth (1 AU away). By 1:00 a.m. (a light-hour), the light would have traveled past Jupiter. By midnight (a light-day), it has traveled 26 Tm, far past the Kuiper belt. But that light would not reach the inner edge of the Oort cloud for another 27 days. It then travels through the cloud for a long, long time. At midnight on December 31, after one light-year, it is still making its way past the icy debris. That same parcel of light wouldn't reach the outer edge of our solar system until the following August, 20 months after it left the sun.

The "year" in "light-year" is the Julian year, which is always exactly 365.25 days of 86,400 seconds each.

Clearly, a light-year is a very long way indeed. In case you're prepping for a game show: a light-year is 9.46 Pm (petameters, that's 10^{15} m); about 6 trillion miles; 63,200 AU; or 31,557,600 light-seconds. For reference, the moon is a bit over 1 light-second from Earth.

But the size of a light-year is truly a matter of perspective, and we'll quickly see that a mere light-year is a measly measurement. After all, to explore the galaxy in which our own solar system belongs, we must rely on kilolight-years (abbreviated kly); traveling between galaxies requires megalight-years (Mly); and supergalactic structures in our universe span gigalight-years (Gly)—literally billions and billions of light-years.

To Infinity . . . and Beyond! On a clear night, in the countryside or from a mountaintop, you can't help but be awed at the sight of kerjillions of stars sparkling overhead. As it turns out, you can actually see only about two thousand stars with the unaided eye from any one spot on Earth, and mostly likely every single one you see is a star in our own galaxy. Until a century ago, astronomers thought that our galaxy was the extent of the universe; as we'll soon see, there's a bit more to the universe than that.

The closest stars to us, deep in the southern sky, are the small cluster of Alpha Centauri A, Alpha Centauri B, and Proxima Centauri, only about 4.2 light-years (about 1.3 parsecs, or 40 petameters) away. To get a sense of the distance, if our sun were the size of a grapefruit in Los Angeles, then Alpha Centauri A would be another, slightly larger, grapefruit in Chicago. If we could send a rocket toward these stars at 80,000 km/h (50,000 mph), it would take 57,000 years to arrive.

This distance between the stars is pretty typical throughout the galaxy, though there are many instances of clusters of two or three stars that are much closer together, like those in Centauri. There is far greater variation in the size of the stars themselves: Some, such as red dwarfs, are much smaller than our own sun, and some are far greater. For comparison, take the red supergiant Betelgeuse—about

> "The cells of our body are as small relative to our own size as a mountain is large."
>
> —Christopher Potter,
> *You Are Here*

640 light-years away and one of the brightest stars in the night sky—in the constellation Orion. With a radius almost 1,200 times larger than our own sun, if you placed Betelgeuse in the center of our solar system, Earth, Mars, and even Jupiter would be inside it.

That's nothing compared with the largest star currently known, VY Canis Majoris, 5,000 ly away from us and twice the size of Betelgeuse. If this star were the size of Mount Everest, our sun would be only 15 feet (4.5 m) wide.

About 27,000 ly from us, in the constellation Sagittarius, there appears to be a supermassive black hole (called Sagittarius A), with a gravitational pull so great that it acts as the centerpoint around which our entire galaxy revolves. We call our galaxy the Milky Way, reflecting the greater density of stars along a cloudy band that streams across the night sky. That river of light is our view of billions of stars spiraling around in a relatively flat plane, about 1,000 ly thick and as much as 100,000 ly (30 kiloparsecs, or 9.5×10^{20} meters) wide. Imagine: If our solar system (out to Pluto) were shrunk to the size of a quarter, our galaxy would be as big as the western half of the United States.

The fact that we can see a similar number of stars on either side of this plane indicates that we are somewhat centered between the "top" and "bottom" of the spiral, with many stars on every side of us. However, if we can see only 2,000 stars unaided, how many stars are really there? Look through the telescope, do the math, and you'll find there are somewhere between 200 and 400 billion stars in the Milky Way, each rotating around Sagittarius A about once every 250 million years.

Residents of the southern hemisphere looking into the nighttime sky may notice two tiny clouds that appear to have broken away from the Milky Way. These are the Magellanic Clouds, first identified by the crew of Ferdinand Magellan's sixteenth-century expedition past South America. More specifically, they are galaxies—the Large Magellanic Cloud (LMC) and the Small Magellanic Cloud (SMC)—and the only objects outside of the Milky Way that we can see without a

"Space is big. Really big. You just won't believe how vastly hugely mind-bogglingly big it is. I mean, you may think it's a long way down the road to the chemist, but that's just peanuts to space."

—Douglas Adams,
The Hitchhiker's Guide to the Galaxy

telescope. Both are dwarf galaxies, less than a seventh the size of our own, with only perhaps 350 million stars between them, and both are more than 160,000 light-years away from us.

Of course, the greater the mass, the larger the gravitational pull—and just as planets are held to stars by gravity, and stars are drawn to black holes in galaxies, whole galaxies are pulled together, too. Those little Magellanic Clouds will likely someday be consumed by the greater collective mass of the Milky Way. But lest we fall pray to braggadocio, we need only look to the Andromeda galaxy, a mere 2.5 million light-years away and containing perhaps as many as a trillion stars. Andromeda appears to be approaching us at 500,000 km/h, and the collision—about two billion years from now—promises to be spectacular.

It might seem that we're starting to get a grasp on the size of the universe, but the Milky Way, Magellanic, and Andromeda galaxies make up only a small part of what's called the Local Group—a small cluster that includes about 30 separate galaxies. The American Museum of Natural History in New York City offers a beautiful reference to the size of the Local Group: In the museum sits the Hayden Sphere, 87 feet (26.5 m) in diameter, and a sign explaining that if the Local Group were shrunk to the size of the sphere, the Milky Way would be about 2.5 feet (80 cm) large. Between those handful of galaxies in the cluster is . . . we don't know, but it's likely pretty much *bubkes*. As the saying goes, "They don't call it *space* for nothing."

The Local Group is, in turn, a tiny piece of the Virgo supercluster, which is about 100 billion times larger than the Milky Way and contains hundreds or even thousands of galaxies. If you don't have vertigo yet, consider this: There are likely around 10 million superclusters in the universe, containing billions (possibly trillions) of galaxies, and likely about 3×10^{22} stars. Astronomers now believe that a large number of these harbor planets, but if even only a tiny percentage have habitable planets, the odds are almost certain that there is sentient life out there, waiting to be found.

The Canis Major dwarf galaxy is just outside our own. In fact, it's actually closer to Earth than is the center of the Milky Way, at only 25,000 ly away.

"There is no smallest among the small and no largest among the large, but always something still smaller and something still larger."

—Anaxagoras, fifth-century Greek philosopher

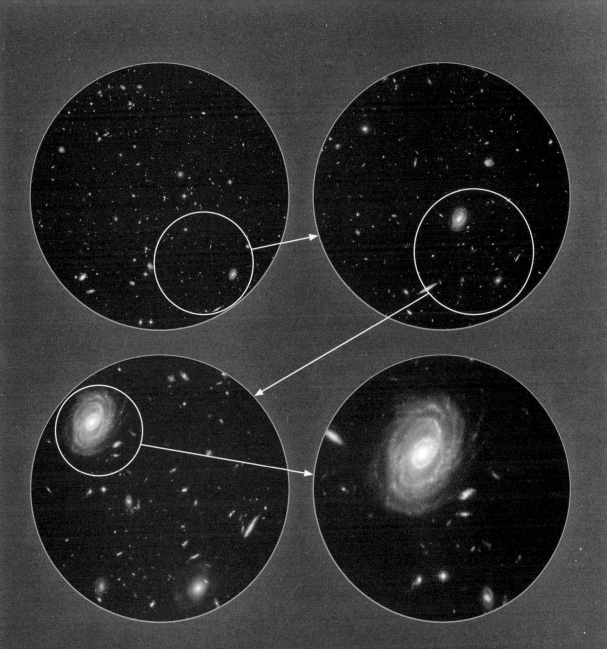

▲ The top-left image, called the Hubble Space Telescope Ultra-Deep Field, represents a span the size of a pen point against the night sky.

When you look up tonight, make a mental note that there are millions of galaxies (that's *galaxies*, not stars) in the "bowl" of the Big Dipper alone.

We cannot see each star in each galaxy, no matter how powerful our telescope. Rather, the lights of whole galaxies blend together to form pinpoints, like the thousands of individual street and house lights of a city coalesce into a single fuzzy spot in a satellite photo. And similar to the patterns of these city lights you can see from space, clusters of galaxies form intricate, cotton-candy-like webs of filaments across the cosmos.

The largest of these filaments is the Sloan Great Wall, discovered in 2003. About 1 billion light-years away, constructed from countless galaxies, and stretching almost 1.5 billion light-years long, the wall is sandwiched between two enormous voids in space. And yet, as unfathomably large as it is, the wall spans only about 1/60 the diameter of the known universe—when you see it against the computer-generated map of all we know about the cosmos,

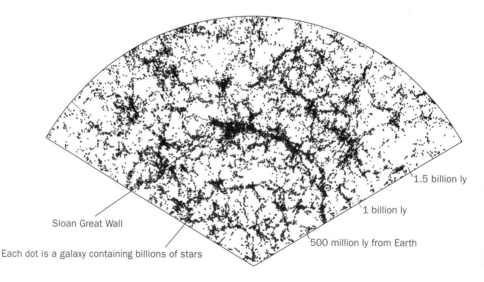

1.5 billion ly

1 billion ly

Sloan Great Wall

500 million ly from Earth

Each dot is a galaxy containing billions of stars

M. Blanton and SDSS Collaboration, www.sdss.org

it's hardly more than a smudge, like something spattered on
a windshield.

Looking out into the universe is like looking through a smoky
room on a sunny afternoon, seeing motes of dust and ash caught in
beams of light. But imagine: Each speck is itself a cluster of galaxies,
separated by unimaginably vast amounts of space. And each galaxy
in that particle is itself a dusty room containing uncountable number
of stars, like ours, the sun.

As far as astronomers can tell, our universe likely has a radius of
about 46 billion light-years (435 yottameters, or 4.35×10^{26} m). This is
far, far larger than it should be if the universe is—as astrophysicists
also believe—"only" 13.7 billion years old. So is the size a fiction? No;
space itself is far stranger than fiction.

Getting Small In contrast to the breathtaking size of the universe is
the world of the very, very small. It's easy to overlook that world; the
very small goes largely unnoticed even though it is every bit as awe-
inspiring. Even the largest superclusters in the cosmos are built from
the same things we are: atoms, subatomic particles, and perhaps
things even tinier than that.

Most of the objects we can hold in our hand fall into the range of
a few centimeters, and the smallest can be measured in millimeters.
This is a level of the human scale we are intimately aware of—we can
see it, even play with it. The smallest mammals, the Etruscan pygmy
shrew and the bumblebee bat, are just over 3 cm long. A penny is
about 1.9 cm wide. The head of a pin is about 1.5 millimeters, and a
grain of salt is about 0.5 mm. You can also write that measurement as
500 micrometers—labeled μm, which is a thousandth of a millimeter,
or a millionth of a meter, and is sometimes still called by its old
name: a micron.

The spectrum of our human vision extends down as far as
about 100 to 200 μm (1or 2 *tenths* of a millimeter). Here we find the
diameter of dust mites and the human ovum (the largest cell in the
human body), or the thickness of a thin human hair or a dollar bill.

However, even though we can see a single grain of sand or a minuscule gnat, it's hard for us to understand that the microworld works differently from our macro expectations. To a tiny insect, the air we breathe and walk through so effortlessly is actually rather thick with molecules—what we would call flying, it would consider swimming. As we then step beyond the seeable, into a world until recently hidden from us, we must begin to learn new sets of rules, ones that become increasingly bizarre from our human perspective.

A glimpse through almost any microscope shows a world where even the smoothest surfaces suddenly become jagged, with unexpected crevices and outcroppings. Like psychedelic fractal artwork, a tiny rock takes on the form of a craggy mountain, and bread mold becomes a field of flowering plants.

Optical microscopes bend light like a set of fun-house mirrors, allowing us to extend our sight below 100 μm, where we find individual animal and plant cells. A paramecium swimming in lake water is as small as 60 μm. Red blood corpuscles are disks about

▼ This moth's eye is 800 μm wide, a hundred times larger than a single red blood cell (bottom right); the large spiked pollen from the Hibiscus flower is about 110 μm wide.

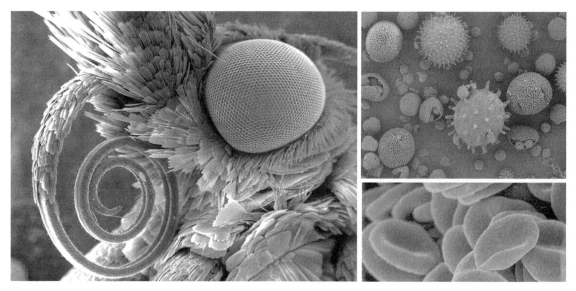

Moth and pollen courtesy Dartmouth University; blood cell courtesy Tina Carvalho

7.7 μm wide and 3.7 μm thick. To get a sense for how tiny this is: If a red blood cell were expanded to the size of an apple, an apple enlarged the same amount would be twice as tall as the Empire State Building.

There are far smaller cells, too: The head of a sperm cell is only about 5 μm long, more than twice the size of that hot-dog-shaped *E. coli* bacterium that thrives in our large intestine. In their world, even pure water is as thick as honey, and astonishingly, about 200,000 of these little creatures could live in the period at the end of this sentence.

With the most powerful optical microscopes, we can magnify objects more than 1,000 times, letting us see organelles inside cells even smaller than a micrometer—things measured in the hundreds of nanometers, or *billionths* of meters. To visualize one billionth of anything, consider half the width of your little fingernail compared with the distance from New York to Los Angeles.

We could make more powerful magnifying lenses, but at this size visible light itself fails us. After all, a single light wave is as small as 400 nm. In order for us to see something, a light wave must be affected by it—either reflected back, refracted away, or absorbed by it. An influenza virus is about 115 nm long, too small to have much effect on a wave of light; a rhinovirus (the predominant cause of the common cold) is even tinier: a small sphere only 20 nm wide.

At this size, we're dealing with tightly packed groups of molecules. And the difference in size between a light wave and a small molecule of atoms is like a 40 m (130 ft) ocean wave encountering a pebble on the beach—the wave may be affected by the entire beach but certainly not by a single rock, or even a pile of them.

Instead, to get a sense of what's going on in the realm of the nanometer, scientists use X-rays, or electrons with wavelengths about 100,000 times shorter than visible light, and watch how their minute electrical charges fluctuate as they interact with molecules. Or, in atomic force microscopy, scientists probe the surface of a material

One billionth: A nanometer is a billionth of a meter; a dime (or 1-yen coin) is about a billionth of the diameter of Earth.

"A cloud is made of billows upon billows upon billows that look like clouds. As you come closer to a cloud you don't get something smooth, but irregularities at a smaller scale."
—Benoit Mandelbrot, mathematician

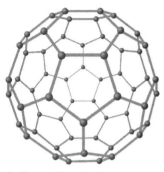

▲ Buckyball molecule

with a sliver of a sliver of a needle, "feeling" the hills and valleys to build a picture of a world beyond seeing.

DNA, that tightly wound double helix of molecular instructions, is only about 2 nm wide—though if you unfurled the coils, the strands of DNA in a single cell would stretch over 2 m (6 ft) long. A single molecule of table sugar (sucrose) is about 1 nm across, as is the soccer-ball-shaped molecule called the buckminsterfullerene or buckyball. Named after the inventor of the geodesic dome and found in tiny quantities in ordinary soot, this molecule is made of 60 carbon atoms fused together, and when compressed it is twice as hard as diamond.

Clap your hands in the "real world," and you can feel the collision generating sound and heat. But in the world of the nanometer—and even smaller, the picometer, which counts out trillionths of a meter—we find that something completely different is happening. Atoms and molecules rarely actually collide (outside of a sun or a supercollider). Instead, they constantly attract or repulse each other, like tiny magnets. As one hand approaches the other, the molecules in your skin strain to hold together and push back against the opposing hand, harder and harder, until no further passage is possible. This whole process is so fast and occurs at such a microscopic level that we are fooled into believing that our hands are solid. Similarly, you may feel your behind against your chair, or your car bumper hitting a concrete wall—the illusion of firm boundaries all around us is compelling, but you're experiencing nothing more than electromagnetism in action.

Atoms are generally a few tenths of a nanometer across, around 100–500 picometers. The smallest atoms (hydrogen, carbon, oxygen, and so on) are all roughly 100 pm in diameter, a measurement that is sometimes called an angstrom (labeled with the Scandinavian Å). They may be magnetically locked together to form solid structures, like crystals; or loosely affiliated in a liquid; or floating free form in a gas, where molecules tend to be well spaced, about 5 to 10 diameters apart.

Once again, it is virtually impossible to truly understand the near-infinitesimal size of atoms, but here's an attempt, using the previously mentioned Hayden Sphere, 87 feet (27 m) in diameter: If a red blood cell were expanded to the size of the sphere—bigger than a six-story building—then a rhinovirus would be about 3 inches (7 cm) in diameter. And if that rhinovirus were expanded to the size of the sphere, then a water molecule would be the size of an inflatable exercise ball, and a single atom would be the size of a basketball.

Or consider this: If you magnified everything so that an apple appeared as large as our Earth, a flea would be as large as a small country, and an amoeba or a human body cell would be the size of a midsized city. A human chromosome would be the size of a baseball field, a virus would fit inside the infield, and a single molecule would fit on home plate.

Dividing the Indivisible Materialists insist that each and every thing is inevitably made of even smaller things. So although the word *atom* by definition (from the Latin *atomus* and the Greek *atomos*, meaning "invisible") implies that there can be nothing smaller, every high school student knows that there's more to be found below the angstrom. On the one hand, you can discover a great deal about the universe inside a single atom; on the other hand, you'll find virtually nothing there.

If you've never been inside the enormous Houston Astrodome, imagine a huge, round sports building, able to seat sixty-five thousand people for a football game. Now imagine a single watermelon seed sitting at the 50-yard line. That tiny pit is like the nucleus of an atom, surrounded by a huge spherical cloud of electrons, far, far away, separated from the nucleus by nothing but space. The size of a hydrogen atom, for example, is about 50,000 times larger than the nucleus itself. Given that the atom (defined by the diameter of that electron cloud) is about 100 pm or 1Å, the nucleus containing a single proton is only a couple of femtometers wide.

OBJECT SIZES

Infinitesimal
String scale
Nano scale
Sub-atomic scale
Atomic
Molecular
Mitochondriatic
Cellular
Microscopic
Minuscule
Tiny
Lilliputian
Small
Medium
Bulky
Large
Immense
Massive
Giant
Mammoth
Colossal
Leviathan
Vast
Galactic
Cosmic
Universal

from Hatch's Order of Magnitude

A femtometer (fm) is really, really small: a millionth of a nanometer—that is, a millionth of a billionth of a meter (10^{-15} m). And it's important to note that words often have different meanings at this subatomic level than in our daily lives. For instance, electrons aren't *things* with size and mass that revolve around the nucleus in the way they're often drawn in diagrams. Rather, each electron is in a constant probability cloud ("it could be here, or here, or here . . ."). Most of the time, in most places, there's no *there* there. No matter how strong our microscopes, we'll never capture more than a blurry photo of an atom, because reality itself at this level is impossible to focus—it's all about probability, not certainty.

▲ Olympicene—named after the Olympic logo—is a single molecule only 1.2 nm wide, made of five interconnected rings of carbon atoms.

What's more, an electron, like a photon of light, is considered an elementary particle—something that truly cannot be broken down into smaller pieces because it does not actually have any size at all. Or, perhaps more accurately, elementary particles do have size, but only sometimes. At other times, they are more like waves of energy, potentialities constantly crossing the threshold between what is and what could be.

The femtometer-sized nucleus—which accounts for 99.9 percent of the mass of the entire atom—is made of one or more protons and neutrons (except for hydrogen, which contains only a proton), and these particles do have mass and can be smashed apart. The result is a menagerie of elementary particles that have individual characteristics and names, but they share a common oddity: Like electrons, they affect space but do not necessarily extend into space as structures with size. Again, at this lowest level, it becomes difficult to discern between energy and matter, so these are sometimes described as being *at most* an attometer (10^{-18}, or a billionth of a billionth of a meter).

Scientists have given these elementary particles the most wonderful names, such as quarks, muons, and leptons, each bonded together with "force carriers" such as gluons and bosons. We know these exist, though several others are only currently hypothesized, such as the tachyon, graviton, and Higgs boson—also called the "God particle" because scientists think it confers mass on other particles, basically turning light to matter.

And just as the American inventor and statesman Benjamin Franklin arbitrarily applied the terms *positive* and *negative* to describe two different types of electric charge, scientists today apply adjectives to the "particle zoo," including *up*, *down*, *top*, *bottom*, *spin*, and *flavor*. For example, a proton is made of two up quarks and one down quark; a neutron is made of one up quark and two down quarks—though try hard enough and you may find a red charm quark, or perhaps even an antiblue antidown quark! The adjectives themselves don't describe the particles—there is no "up" or "down"

If Earth were the size of a baseball sitting at home plate in a baseball park, the moon would be the size of a cherry, about 7.5 feet away. Mars is only about a third of a mile away. The sun would be 27 feet across, three quarters of a mile away. The next planet out, Jupiter, would be about 3 miles away. At this scale, the nearest star outside our solar system would be located far off any map of the world.

If Earth were the size of a grain of sand in San Francisco, the nearest star (apart from our own sun) would be the size of a peppercorn, hundreds of miles away, near the Grand Canyon. Sirius, the brightest star in the night sky, is as big as a baseball, halfway across the United States.

or visible color at this level—but they're helpful in distinguishing one category from another.

So if it stops making sense to apply size at this subatomic level, why continue? Isn't delving deeper like bringing a footstool to the peak of Everest to stand on, just to say you went a little higher? For better or worse, scientists are, if nothing else, driven by insatiable curiosity—the word *science* itself derives from the Latin word meaning to know, or to separate one thing from another. So might quarks and other astonishingly small particles such as the elusive neutrino be made of something finer?

Here it all just becomes theory, though theory based on extraordinary research and consideration. The leading idea is that underneath it all, everything is made of "vibrating strings" in an 11-dimensional universe. These strings are about 1.6×10^{-35} meters long, a size called the Planck length. In other words, compare the size of a single atom to the length of your arm; that's about how much smaller a string would be compared to an entire atom. Or imagine: If you magnified a single atom to the size of our entire solar system, one Planck length would be the width of a strand of DNA.

The Planck length also marks the smallest measurement that makes any sense. That is, given the speed of light, the force of gravity, and other universal constants, physicists have calculated that nothing *can* be smaller. If you think of our reality as being created out of tiny squares, like pixels on a computer screen, then each pixel is 1 Planck length tall and wide. We simply cannot venture smaller.

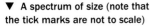

▼ A spectrum of size (note that the tick marks are not to scale)

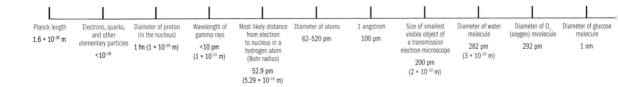

Planck length	Electrons, quarks, and other elementary particles	Diameter of proton (in the nucleus)	Wavelength of gamma rays	Most likely distance from electron to nucleus in a hydrogen atom (Bohr radius)	Diameter of atoms	1 angstrom	Size of smallest visible object of a transmission electron microscope	Diameter of water molecule	Diameter of O_2 (oxygen) mvolecule	Diameter of glucose molecule
1.6×10^{-35} m	$<10^{-18}$	1 fm $(1 \times 10^{-15}$ m$)$	<10 pm $(1 \times 10^{-11}$ m$)$	52.9 pm $(5.29 \times 10^{-11}$ m$)$	62–520 pm	100 pm	200 pm $(2 \times 10^{-10}$ m$)$	282 pm $(3 \times 10^{-10}$ m$)$	292 pm	1 nm

Size Depends on Space Unfortunately, there is a fundamental problem with any discussion of the spectrum of size and dimension: Size depends on space; that is, every measurement is based on how much space (length, width, and height) something takes up. And—as weird as this may sound—scientists still don't understand what space is or how it works.

Everyone knows that science and math go hand in hand, but few people understand the extent to which scientists and mathematicians rely on philosophy to get the job done. As much as we want to believe that science teaches absolute truth, the absolute truth is that science is based on assumptions and hypotheses, and in some cases we may simply not be able to prove that some of those assumptions are valid. This is perhaps never more true than when discussing space.

The brilliant physicist Isaac Newton made his opinions clear in his late-seventeenth-century opus, the *Principia Mathematica*: Space and time are absolutes, a standard in which all things have their place and order. Newton's firm grip on reality—with its rigid, invisible scaffolding that gives the cosmos its shape—is comforting. In Newton's world, a ruler is a ruler is a ruler—the very essence of modernism. But of course, we're also talking about a guy who, in the name of science, stuck a blunt needle between his own eye and ocular bone just to see what was back there.

While Newton was probing his absolute universe, the mathematician Gottfried Leibniz was arguing that everything in our

▲ Sir Isaac Newton

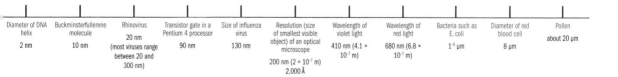

Diameter of DNA helix	Buckminsterfullerene molecule	Rhinovirus	Transistor gate in a Pentium 4 processor	Size of influenza virus	Resolution (size of smallest visible object) of an optical microscope	Wavelength of violet light	Wavelength of red light	Bacteria such as E. coli	Diameter of red blood cell	Pollen
2 nm	10 nm	20 nm (most viruses range between 20 and 300 nm)	90 nm	130 nm	200 nm (2 × 10⁻⁷ m) 2,000 Å	410 nm (4.1 × 10⁻⁷ m)	680 nm (6.8 × 10⁻⁷ m)	1⁻⁵ μm	8 μm	about 20 μm

▲ Gottfried Leibniz

universe is positioned and moves relative to everything else. That is, objects don't exist inside a fixed space; space itself is defined as the relation between the objects.

This may seem like nitpicking, but these underlying assumptions about what space is turn out to have radical implications on science and how we measure things. For example, the data that astronomers collect from distant stars would have to be interpreted completely differently with each model, leading to very different understandings of the cosmos.

However, in the early twentieth century, Einstein's theory of relativity prompted a complete rethinking of the matter. Space, it turns out, *is* based on your frame of reference, and, what's more, it's far from absolute; rather, it is warped by mass and motion. For example, an object traveling faster becomes shorter and heavier, but only compared to one moving more slowly—that is, it's relative. Here's another oddity: The more massive an object, the more it can literally warp space and time, like a block of spongy foam twisted and stretched. From any one point inside space it doesn't appear warped, but careful measurements of light moving through space expose the truth. Einstein's theories have since been repeatedly validated with experimental data: The fabric of space is elastic, not concrete.

Based on Einstein's descriptions, the physicist Hans Reichenbach wrote *The Philosophy of Space and Time* (1928), in which he pointed out that you cannot know the inherent, absolute size of an object; you can know it only relative to another object. Of course, if everything is relative, including time and space, we end up in a vicious circle: How

Length of sperm cell (head to tail)	Size of smallest dust particles	Thickness of a dollar bill or average human hair	Largest cell in human body; ovum	Dust mite	Dot at end of a sentence	Grain of salt	Length of jump of common flea, 1.5 mm (0.06 in) long	Diameter of the head of an average pin	Height of a line of text in 12-point type	Length of smallest vertebrate
$50\ \mu m\ (5 \times 10^{-5}\ m)$	0.1 mm	110 μm	140 μm	200 μm	300–500 μm	0.5 mm	330 mm (13 in.)	1.7 mm	4.234 mm	*Paedophryne amauensis* frogs of
	$100\ \mu m\ (1 \times 10^{-4}\ m)$	$(1.1 \times 10^{-4}\ m)$; 0.0043 in., 0.11 mm				(500 microns)		$(1.7 \times 10^{-3}\ m)$	$(4.234 \times 10^{-3}\ m)$	Papua New Guinea
										7.5 mm

can we understand the underlying forces of nature if we can't get a grip on the geometry of space-time? And how can we understand the geometry of space-time if we don't understand the underlying forces? "It appears," wrote Reichenbach, "that the solution of the problem of time and space is reserved to philosophers who, like Leibniz, are mathematicians, or to mathematicians who, like Einstein, are philosophers."

Ultimately, we humans experience our universe like the proverbial blind men and the elephant, where different perspectives result in very different "truths." We start by understanding things within our reach—on our human scale—and at this level Newton's laws generally work. Then, as we extend our reach by using instruments, we gather data that makes no sense from our original perspective; it's as though the rules change in the worlds of the very large and the very small.

For example, a traditional model says that stars and galaxies are all flying farther away, like the detritus of some massive explosion. But current theories actually point to something else: that space itself—the "nothingness" between these massive bodies—is stretching, like the surface of a balloon being blown up, or expanding, like bread rising in an oven, largely with the help of what cosmologists call dark energy. (This should not be confused with dark matter, which is, no pun intended, a different matter entirely.)

As far as we can tell, space is expanding at about 70 kilometers per second per megaparsec (3.2 light-years). In other words, let's say you could stretch a tape measure out to Proxima Centauri. You could look down at the number on the tape to see just how far away

▲ Albert Einstein

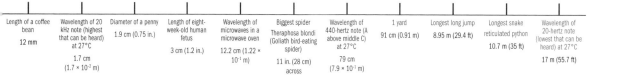

Length of a coffee bean	Wavelength of 20 kHz note (highest that can be heard) at 27°C	Diameter of a penny	Length of eight-week-old human fetus	Wavelength of microwaves in a microwave oven	Biggest spider Theraphosa blondi (Goliath bird-eating spider)	Wavelength of 440-hertz note (A above middle C) at 27°C	1 yard	Longest long jump	Longest snake reticulated python	Wavelength of 20-hertz note (lowest that can be heard) at 27°C
12 mm	1.7 cm (1.7 × 10⁻² m)	1.9 cm (0.75 in.)	3 cm (1.2 in.)	12.2 cm (1.22 × 10⁻¹ m)	11 in. (28 cm) across	79 cm (7.9 × 10⁻¹ m)	91 cm (0.91 m)	8.95 m (29.4 ft)	10.7 m (35 ft)	17 m (55.7 ft)

it was—but when you brought the tape measure back in, you'd see that the distances between the marks had literally changed; the tape measure had actually stretched, invalidating the measurement you had just taken.

Astronomers see the effects from space-time expansion every day, as light emitted from far-off superclusters of galaxies travels through unimaginable amounts of space to reach us. As space stretches, the light waves themselves also become elongated, causing the color to shift toward the red end of the spectrum (called "redshift").

The fact that space is stretching leads us to another astonishing possibility: Extremely distant objects may be moving away from us at a speed greater than the speed of light. The objects themselves aren't breaking any speed laws, but the cumulative effect of the expansion of space adds up. If this is so, then there may be far more to our universe than we can observe—it would be impossible for light from beyond that far horizon to ever reach us.

Innerspace When we take our eyes off the stars for a moment, and turn toward the atomic and subatomic worlds, similarly bizarre effects await us. Remember that all material is made of molecules, which are made of atoms; that the atoms don't even touch each other but are held together by electromagnetic forces; and that the atoms themselves are virtually all just space.

If you closely inspect a color photograph printed in a newspaper or magazine, you'll find that, at its heart, the vibrant and compelling images are all constructed from tiny dots arrayed on a grid. Even the rich spectrum of color is a trick, as the spots are printed using

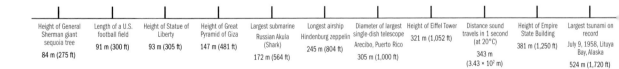

| Height of General Sherman giant sequoia tree
84 m (275 ft) | Length of a U.S. football field
91 m (300 ft) | Height of Statue of Liberty
93 m (305 ft) | Height of Great Pyramid of Giza
147 m (481 ft) | Largest submarine Russian Akula (Shark)
172 m (564 ft) | Longest airship Hindenburg zeppelin
245 m (804 ft) | Diameter of largest single-dish telescope Arecibo, Puerto Rico
305 m (1,000 ft) | Height of Eiffel Tower
321 m (1,052 ft) | Distance sound travels in 1 second (at 20°C)
343 m
(3.43 × 10² m) | Height of Empire State Building
381 m (1,250 ft) | Largest tsunami on record
July 9, 1958, Lituya Bay, Alaska
524 m (1,720 ft) |

only four pigments. From a distance, it all blurs together to convey a picture. But is what we call reality any different? Is it not just a series of dots—matter loosely held together—that we interpret as Truth?

And just as the ink can't perform its magic without paper, atoms cannot manifest without space. Space itself—which we ignore, like the black-clad puppeteers in a Japanese Bunraku play—turns out to be a medium. The spaces between are not empty, after all. Rather, space is woven tightly with a tempest of electrons, gluons, photons, bosons, neutrinos, waves of probability, fields of potential. And these elements play by rules we don't yet understand. For example, in certain circumstances particles can become entangled in such a way that they behave as a single entity, no matter how far apart they are. This has been repeatedly demonstrated, even though it completely violates the rules of classical physics, leading Einstein to famously call it *spukhafte Fernwirkung*, or "spooky action at a distance."

Then, as we peer closer, far down below the attometer, the fabric of space-time loses its smooth peculiar-but-reliable uniformity and becomes unruly. Like driving out of range of an analog radio station, the signal is slowly replaced by static. Get small enough and probability gives way entirely to randomness, so that physicists now believe that, at the Planck scale, space is a foamy, frothy sea of possibility. At this quantum level, virtual particles form out of energy and dissolve almost instantly, black holes may suddenly pop in and out of existence, and wormholes leading from one part of the universe to another (or possibly even into other universes) may appear then disappear in an instant.

Height of Burj Khalifa	Tallest waterfall Angel Falls in Venezuela	1 mile	Depth of Lake Baikal, deepest freshwater lake	Average depth of world's oceans	Height of Mt. Everest	Depth of Marianas Trench in the Pacific Ocean	Height of the troposphere at equator	Average altitude of International Space Station above sea level	Length of Grand Canyon	Diameter of moon
29.84 m (2,723 ft)	979 m (3,212 ft)	1.61 km (5,280 ft)	1,620 m (5315 ft)	3,790 m (12,434 ft)	8,848 m (29,028 ft)	10,918 m (35,820 ft)	17 km (56,000 ft)	370 km (230 mi)	446 km (277 mi)	3.47 × 10⁶ m

Of course, once again, nobody knows if any of this is really How It Is. We've built myriad mathematical constructs that attempt to describe our universe—some that are completely at odds with each other—and discovered that there may be more than one correct answer, depending upon your perspective. While most people imagine the universe to be roughly spherical, it turns out that space-time may actually be the shape of a hyperbolic saddle, or perhaps (the current forerunner idea) based on a buckyball-shaped Poincaré dodecahedron. Superstring theory, a sweet mix of philosophy and mathematics, tells us that space may contain five or six additional, infinitesimally small dimensions that are twisted in intricately folded shapes called Calabi–Yau manifolds.

The math clearly tells us that the universe should contain gravitational waves that literally stretch and compress space, but in reality the effects are so minute that no one has yet detected them. One difficulty is that noise creeps into the experimental data—a meaningless static that obscures our view, like poor reception on an old rabbit-eared television. A reasonable (though controversial) explanation is that we are actually seeing evidence of Planck-length froth at much larger sizes, and that we're seeing it because—sit down for this one—the universe as we know it may actually be a hologram based on some far more complex reality played out on an insanely large membrane, like the way we see a three-dimensional image on a flat credit card. Could it be that, once again, our fundamental assumptions are based on illusions?

> **You know about measuring in** three dimensions: length, width, and breadth (or height). But many scientists envision a fourth spatial dimension—sometimes called ana/kata. A 4-D cube is called a tesseract.

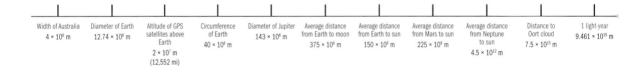

Width of Australia	Diameter of Earth	Altitude of GPS satellites above Earth	Circumference of Earth	Diameter of Jupiter	Average distance from Earth to moon	Average distance from Earth to sun	Average distance from Mars to sun	Average distance from Neptune to sun	Distance to Oort cloud	1 light-year
4×10^6 m	12.74×10^6 m	2×10^7 m (12,552 mi)	40×10^6 m	143×10^6 m	375×10^6 m	150×10^9 m	225×10^9 m	4.5×10^{12} m	7.5×10^{15} m	9.461×10^{15} m

As we go looking for answers, we should keep in mind the words of the twentieth-century British mathematician and philosopher Alfred North Whitehead: "There are no whole truths; all truths are half-truths. It is trying to treat them as whole truths that plays the devil."

Stretching Beyond *Homo sapiens* have been around for 100,000 years or so, but only in the past few hundred years have we had even a vague sense of the universe—big and small. Recognizing the orbiting planets led to theories of the stars; other explorers looked in the other direction, guessing that there was such a thing as microscopic life. But it wasn't until a century ago that we found the multitude of galaxies and uncovered the truth of subatomic particles.

Each discovery has led us to new mysteries; each mystery has led to new ideas. And at every step, there have been some who declare that now, finally, we understand, and others who understand that only now, finally, we can learn more. As the science writer Isaac Asimov wrote, "We have been misled before, and it may be that in time to come, additional vastness and intricacy will unfold and we will come to realize that what we now know, or think we know, is but a tiny part of a still greater whole."

Distance to Alpha Centauri A	Width of Milky Way galaxy	Distance to Andromeda galaxy	Diameter of Local Group (few dozen galaxies)	Diameter of Virgo Supercluster (1,000+ galaxies)	Length of Sloan Great Wall	Distance to edge of the observable universe
4.3 ly (4×10^{16} m)	100,000 ly (9.5×10^{20} m)	2.5 million ly (2.4×10^{22} m)	10 million ly (9.5×10^{22} m)	110 million ly (1×10^{24} m)	1.37 billion ly (1.30×10^{25} m)	46 billion ly (4.35×10^{26} m)

LIGHT

What is to give light must endure burning.

—Viktor Frankl

WHAT IF YOU COULD WATCH MUSIC AS IT STREAMED OUT FROM A radio tower, like an enormous lightbulb shining in the sky? What if turning on your television made your eyes blink from the bright flash—not from the screen but from the tip of your remote control? What if your microwave heated your food with light, like one of those old toy ovens from your childhood? In fact, all these things— radio, cell phones, microwave ovens, and remote controls—are based on light. They use light that, even though we cannot see it, is nevertheless the same in every way as the light that we can see.

We are constantly awash in an astonishing spectrum of light, everflowing and everlasting. Even in the darkest room we cannot escape light, if only because our own bodies radiate it through the very act of living.

Of course, humans are mercifully sensitive to only a small portion—less than a thousandth of 1 percent—of the full spectrum of light. Our eyes see the edges of a rainbow fade gradually away to what seems like nothingness. But electronic instruments uncover for us a world far beyond the red (on one side) and violet (on the other). The universe truly is far stranger (and brighter!) than we can imagine.

Such Stuff as Light Is Made On Light—that is, the light that the physiology of our eyes is tuned to see—is part of a phenomenon called electromagnetic radiation (or EMR) that describes how electricity and magnetism radiate, or travel, from one place to

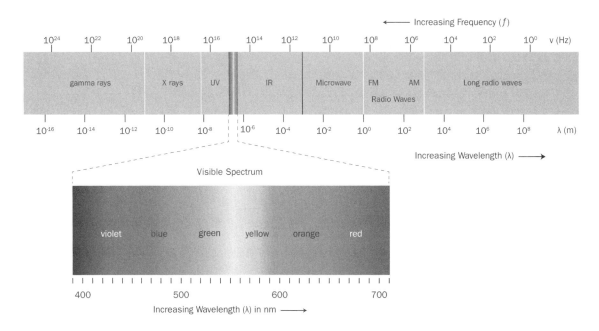

Increasing Frequency (*f*)

| 10^{24} | 10^{22} | 10^{20} | 10^{18} | 10^{16} | 10^{14} | 10^{12} | 10^{10} | 10^{8} | 10^{6} | 10^{4} | 10^{2} | 10^{0} | v (Hz) |

gamma rays   X rays   UV   IR   Microwave   FM   AM   Long radio waves

Radio Waves

| 10^{-16} | 10^{-14} | 10^{-12} | 10^{-10} | 10^{-8} | 10^{-6} | 10^{-4} | 10^{-2} | 10^{0} | 10^{2} | 10^{4} | 10^{6} | 10^{8} | λ (m) |

Increasing Wavelength (λ) ⟶

Visible Spectrum

violet   blue   green   yellow   orange   red

400   500   600   700

Increasing Wavelength (λ) in nm ⟶

▲ **Visible light spectrum**

another. Understanding EMR is necessarily technical, but the more you understand about light, the more amazing it is.

When an electrical current flows from one place to another, such as through a wire or the nerve system in your body, it generates a magnetic field. That's why an electrical current flowing near a compass moves the needle. It's why your favorite tune can be converted from an electrical pulse in a wire, to a magnet in a speaker, to the sound you hear. Control the current, and you control the magnet, which controls the speaker vibrations we hear as sound.

Conversely, when you move a magnet near a coil of wire, it creates electricity in that wire. This intimate bond between electricity and magnetism is what makes it possible for us to create motors, or, in fact, to create electricity itself in a generator.

So a changing electrical current creates a magnetic field, and a changing magnetic field creates an electrical field. And an amazing thing happens if you oscillate them by varying these fields back and forth: You create a wave effect, where the changing electric field actually generates the magnetic field, which creates the next electric field, and so on, potentially forever.

This self-propagating "electromagnetic" wave is what we call "light," and it can travel through the vacuum of space without slowing, without fading. This explains why the burning plasma of a distant star, aglow with the dance of free-floating electrons, can throw its radiation waves out across trillions of kilometers to be captured by our eyes and instruments.

Electromagnetic radiation—where energy is moved through space as light enables us not only to see our own sun but also to feel its warmth and, indeed, be burned by it, even on a cloudy day. And it allows us to send and receive seemingly invisible messages with satellites in orbit or with a cell phone tower on a hill.

Making Light Technically, there are two ways to create light: incandescence and luminescence. The former comes from heating a material, such as the nuclear reactions in the sun, or from electricity being passed through a tiny wire filament inside an incandescent lightbulb until it glows white-hot at over 2,000°F.

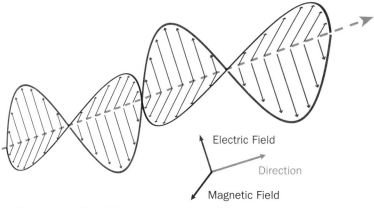

Electric Field

Direction

Magnetic Field

▲ Electromagnetic radiation

Luminescence is light without heat. For example, a watch dial painted with a phosphorescent substance will absorb energy in a lit room and then gently continue to glow in the dark. A firefly generates bioluminescence with a chemical reaction inside its abdomen. The phosphor in some laundry detergents is called a "brightener"—making white clothes "whiter"—because it literally glows with cool, visible light when energy from the sun hits it.

You can see this effect most clearly in a disco or an amusement park fun house that has black lights—special lightbulbs that emit ultraviolet radiation and make the phosphorescent materials in the room glow. Laundered white shirts and socks, natural phosphors in your teeth, and fluorescent paint all pop out brightly because they respond to the energy by giving off light we can see.

If you paint the inside of one of these lights with a phosphorescent coating, the surface itself glows white instead of your T-shirt. The result is the most common form of luminescence we see around us: a plain fluorescent lightbulb.

Both incandescence and luminescence derive from the same underlying mechanism: Electrons in an atom absorb energy of some sort and then emit it in the form of a bit of electromagnetic radiation. In incandescence, atoms are heated until their electrons get so excited that they break free and travel from one atom to another before settling down, releasing their energy as light. In luminescence, electrons are energized but remain within their atoms, quickly emitting the absorbed energy and dropping back to their original level. In both cases, the process continues until you stop adding energy, flicking off the switch.

How Fast, How Red Of course, there are many kinds of light: blue, infrared, ultraviolet . . . and the difference? It's the result of wavelength. Like ripples on a pond, or waves in an ocean, light waves radiate with energy, created by the constant rise and fall of electric and magnetic fields. If the waves gently undulate back and forth, we say they have a long wavelength and a low frequency. That is, it takes

▼ High frequency, small wavelength, more power

▲ Low frequency, long wavelength, less power

more time for each long wave to crest, so the frequency of those waves is slower.

On the other hand, a fast wave—one that goes from electric to magnetic, crest to trough, quickly—has a very short wavelength and a fast frequency.

Frequency and wavelength are always inextricably tied because of a crucial universal constant: the speed of light. This law tells us that there is one speed at which all light—indeed all electromagnetic radiation—travels in a vacuum. It's important to add the vacuum caveat because light does travel slightly slower in a gas, liquid, or clear solid. Place the tip of a stick in a clear brook and you'll see it "bend" underwater because the speed of light is literally slower beneath the surface than it is in air.

Pass light through a diamond and it slows to about 40 percent its normal speed. In fact, through some exceedingly clever tricks involving shooting lasers into extremely cold clouds of rubidium and helium gas, scientists have even been able to slow a beam of light until it is virtually standing still, apparently extinguished, but actually just on pause.

But in the vacuum of space, light travels at about 300,000 kilometers per second (just over 186,200 miles per second). Nothing travels faster.

So if light is radiating out at, well, the speed of light, then the number of waves that pass by a given point in space each second (frequency) is always tied to how long those waves are. Longer wave, fewer of them can go by each second; short little wave, you can cram a bunch of them in each second.

Note that because the speed of light is so (unbelievably, shockingly) fast, when you look at wavelength and frequency, you have to deal with some really huge (and small) numbers.

Red light—that is, light that our human eyes perceive as red—has a frequency of about 420 THz (teraherz). That means those electric fields and magnetic fields are flipping back and forth about 420 trillion times each second. If your car wheels revolved that quickly, you could drive from one end of our solar system to the other in the time it takes to blink an eye.

At that rate, each of those red-colored light waves is about 700 nm (700 billionths of a meter) long, about a tenth the size of a red blood cell, even smaller than a typical microscopic bacterium.

As you shorten the wavelength—that is, increase the frequency, so you get more waves per second—red becomes yellow, then green, then blue, and then, at about twice the frequency of red, purple. Speed up the frequency even more and the light changes, moving beyond what we can see into the ultraviolet (UV) range, then X-rays, then gamma rays. More about those bad boys in a minute.

> "Come forth into the light of things,
> Let Nature be your teacher."
>
> —William Wordsworth,
> "An Evening Scene on the Same Subject"

> "Colours seen by candle-light
> Will not look the same by day."
>
> —Elizabeth Barrett Browning,
> "The Lady's 'Yes'"

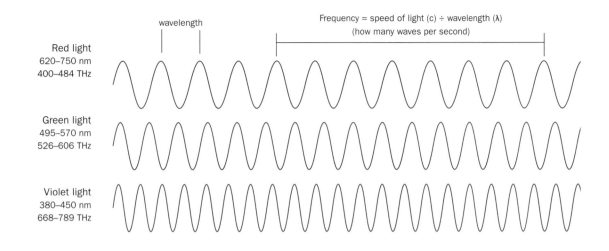

wavelength

Frequency = speed of light (c) ÷ wavelength (λ)
(how many waves per second)

Red light
620–750 nm
400–484 THz

Green light
495–570 nm
526–606 THz

Violet light
380–450 nm
668–789 THz

On the other side of the spectrum, if you slow the frequency (stretching out the wavelength), your red light becomes invisible infrared ("below red"), then microwaves, and then radio waves. The electromagnetic waves of your favorite FM radio station are radiating only at somewhere between 88 and 107 MHz (megahertz, or million cycles per second). Do the math and you'll find that each of those musical waves is about 3 meters (9.8 ft) long. That's longer than you might expect, but remember that each of those light waves speeds by unbelievably quickly—if radio signals could bend, they could travel around the globe 7 times each second.

We know of no longest wavelength in the universe. On Earth, telluric currents—extraordinarily subtle shifts in electromagnetic waves within Earth's crust or oceans, often due to weather or even the silent interaction between the solar wind and Earth's magnetosphere—may be as slow as a few hundred cycles per second. Waves at that rate are called extremely low frequency (ELF). Even longer, in the field of ultra low frequency, are the intricate but slow-flipping waves of our own brains. Each message that fires through

"It is sometimes said that scientists are unromantic, that their passion to figure out robs the world of beauty and mystery. But . . . it does no harm to the romance of the sunset to know a little bit about it."

—Carl Sagan

the brain's electrochemical connections has an electromagnetic result—we are each walking transmitters, though fortunately our signals are weak.

As the frequency drops even lower, to a single cycle each second (1 Hz), such as can be found deep in the movements of Earth itself, the wavelength becomes stretched out across 300 million meters—about 80 percent of the way from Earth to the moon.

It's All Energy As light radiates, it carries energy. You might even say that light is energy in motion. And the amount of energy transmitted through light is based entirely on the light's frequency or wavelength. The higher the frequency, the more energy. In other words, visible light contains more energy than infrared or radio waves. However, more energy doesn't necessarily result in effects you might expect.

While the light we can see will illuminate us, it doesn't warm us much. Most visible light bounces off our skin or is absorbed in ways that don't translate into heat. On the other hand, we are warmed by infrared light; it's invisible and transmits less energy, but it penetrates deeper, and is quickly absorbed by many materials, causing them to warm up.

Lower the frequency even more and you get microwaves, which can melt butter in seconds! Turn on a microwave oven and you send invisible electromagnetic radiation through the air to your food. This light energy has a frequency of 2.45 GHz (gigahertz, or billions of cycles per second)—lower than infrared but still far higher than radio waves. The choice of frequency results in some interesting effects: Microwaves at these frequencies are absorbed by particular kinds of molecules, such as those in water, fats, and sugars, causing them to vibrate, bang into each other, and heat up. The waves just pass by other molecules, such as the dry exterior of a kernel of popcorn or a potato, leading people to exclaim, erroneously, that microwaves cook from the inside out.

> **"To gaze is to think."**
> —Salvador Dalí

	Radio	Microwave	Infrared	Visible light	Ultraviolet	X-ray	Gamma ray
Frequency	30 kHz–300 MHz	300 MHz–3,000 GHz	3–400 THz	400–790 THz	790 THz–30 PHz	30 PHz–3 EHz	3 EHz–3 ZHz (and beyond)
Wavelength	1 Mm–1 m	1 mm–100 μm	750 nm–100 μm	750–400 nm	400–10 nm	10 nm–100 pm	<100 pm
Energy	128 peV–1.25 μeV	1.25 μeV–12 meV	12 meV–3 eV	1.6–3 eV	1.6–3 eV	120 eV–12 keV	12 keV+

▲ Electromagnetic radiation (EMR or "light"). Note: There are no strict boundaries between one type of radiation and another, so these numbers are all approximations.

If you worry about those microwaves escaping the oven, fret not: Waves at this frequency are about 122 millimeters (4.8 in) long—far larger than those of visible light. So the visible light waves can escape through those tiny holes in the glass door, letting us see our food being cooked, while the long microwaves remain inside, bouncing around the chamber.

When you hear that microwaves are electromagnetic radiation, you might get nervous about that word: *radiation*. After all, it's touching stuff you're going to put in your mouth. Fortunately, EMR at the energy of visible light and lower is safe because it's *non-ionizing*. Ionizing radiation in high-frequency, tiny-wavelength light (from ultraviolet light to X-rays and gamma rays) packs such an energy wallop that it can knock electrons right out of their atoms, making for an unstable chemical situation. These reactive atoms are called free radicals and are infamous for their destructive effects.

It's the infrared that makes you sweat on a hot day, but it's the ultraviolet that can give you a sunburn or worse. In the upper atmosphere, seemingly innocuous chlorofluorocarbons, escapees from old refrigerators and air conditioners, are bombarded with ionizing ultraviolet light from the sun. As electrons are blasted off, the newly radicalized molecules do their best to destroy our planet's ozone layer.

It's the ozone layer, of course, that has traditionally helped block much of that same dangerous light from getting down to us. But inevitably some high-powered ultraviolet light pushes through anyway, striking our skin hard enough to break down the fundamental building blocks that keep us alive. When atoms in our DNA are radicalized, the result can be mutant or cancerous cells,

"A 60 W tungsten bulb, a normal household bulb, consumes more than six times the electrical power of a 9 W compact fluorescent lamp but they are both perceived as producing approximately equal amounts of light . . . This is because a lot of the power used by a tungsten bulb is given out in the infrared part of the spectrum where the eye has no response. The light given out by the fluorescent lamp corresponds more closely to the peak sensitivity of the eye."

—UK National Physical Laboratory

so damaged that our internal defense mechanisms can no longer repair them.

You can run inside, but you can't hide. Glass reflects, absorbs, or scatters about 37 percent of low-powered ultraviolet light (called UVA)—that's about as good as putting on sunscreen, but you can still get a sunburn inside your car on a long drive. Small doses of the more powerful (higher-frequency) UVB rays are good for us—they get our body to produce essential vitamin D, but too much exposure is a major cause of melanoma.

Scientists measure the energy in light by electron volts (eV). For example, the light we can see contains only 1 or 2 eV. At 3 or 4 eV, light becomes ionizing. When the light waves speed up to about 30 PHz (petahertz, or 3×10^{16} cycles per second), they carry charges in the hundreds of electron volts. This kind of light has such peculiar properties that the researchers who first discovered them labeled them "X" rays. So powerful, with wavelengths so small, X-rays can slip between molecules of soft material like our skin and organs, stopping only when they encounter dense material such as metal or bone.

> **"Light is not so much something that reveals, as it is itself the revelation."**
>
> —James Turrell, artist

Sure, X-rays leave a path of destruction in their wake, but with short, infrequent doses the risk is low and your body generally repairs minor damage. Most of the X-rays we encounter in our lives are natural: exposure to tiny bits of radioactive rocks in the earth, radiation from the sun, and so on. Again, score one for our upper atmosphere, protecting us from the worst of it.

But as the energy in light increases, so does the danger. When you boost the energy into the tens of thousands of electron volts, the light wavelengths are reduced to picometers—trillionths of a meter, even smaller than atoms. This is the realm of gamma rays.

Gamma rays are just another form of light, but with electromagnetic frequencies measured in exahertz, over a thousand times greater than some X-rays and a million times greater than visible light. Doctors can shine gamma rays emitted from radioactive materials on our body to create incredibly detailed pictures of what's going on inside tissue or bone, or concentrate these rays on an area of cancerous cells to destroy them. Customs officials can bombard shipping containers with gamma rays to "look" through 18 cm (7 in.) of steel and find stowaways or contraband inside.

Like X-rays, gamma rays naturally occur all around us in tiny amounts. When government officials attempt to locate nuclear material with gamma ray detectors, they're often stymied by a wide variety of food. Bananas and Brazil nuts, for example, tend to have higher than average quantities of naturally occurring radioactive material, causing false positives for investigators. It gives new meaning to getting high energy from eating fruits and nuts.

Of course, while these ubiquitous gamma rays are powerful, they're paltry compared to the far end of the electromagnetic spectrum, where gamma rays carrying 20 million electron volts flash from the tops of thunderclouds during lightning storms here on Earth. Traveling outside our terrestrial bubble, the most energetic phenomena in the universe—black holes and supernova—blast out gamma rays that radiate at 10^{27} Hz and more than 5 trillion electron volts.

If these numbers seem extraordinary, it's worth putting them in perspective. It takes 1 joule* of energy to lift an apple off a table, and 1 joule is about 6 exa-electron volts—6 *billion* billion (6×10^{18}) eV. In other words, light is extremely powerful . . . to extremely small things. Light can shatter the infinitesimal world of subatomic particles, but it exerts hardly any pressure on our everyday "macro" world.

That said, with enough light you can achieve the seemingly impossible. Proof came in 2010 with the launch of the Japanese spacecraft IKAROS (Interplanetary Kite-craft Accelerated by Radiation Of the Sun). Outfitted with a microthin solar sail 14 m (45 ft) wide filled with pressure from gentle waves of light, this lightweight unmanned vessel slowly, methodically gained momentum outside

*A joule is a measurement that describes energy or work.

"Every man takes the limits of his own field of vision for the limits of the world."

—Arthur Schopenhauer

The sun is like a 4×10^{26} watt lightbulb. That's brighter than the average star, but there are far brighter ones out there. Epsilon Orionis (the middle star of Orion's belt) is 1,300 light-years away and 400,000 times as bright as the sun. There's a star in the Large Magellanic Cloud called R136a1 that is as bright as almost 9 million suns. Of course exploding stars, called supernovae, are even brighter; the brightest on record peaked at about 100 billion suns.

the drag of Earth's gravity, like Aesop's tortoise. Once considered the stuff of science fiction, IKAROS has already sailed past Venus on the force of light, and solar sailing is thought by many to be the future of interplanetary travel.

> "In the beginning there was nothing. God said, 'Let there be light!' And there was light. There was still nothing, but you could see it a whole lot better."
> —Ellen DeGeneres

A Discrete, Not a Continuous Spectrum So far we've been discussing light as though it were simply "wave energy," but the bigger picture is far more weird.

If you reduce the energy of light, you would expect a smooth continuum, smaller and smaller, like turning down the volume on a stereo until you finally hit zero. But light doesn't work like that. It turns out that light can exist only at particular energy levels—as though the volume knob had notches in it at 3, 2, 1, and 0, nothing between.

The only reasonable explanation for this is that light, at its core, is made of particles. A single particle of light, called a photon, is like a little packet with a discrete amount of energy.

The intensity of light involves the number of photons, but the energy of the photons is something completely different. Imagine a photon as a Ping-Pong ball. If the ball is moving slowly when it hits you, you barely feel it. Now let's increase the intensity by throwing 100 slow-moving Ping-Pong balls at you. More pressure, but because each ball is low powered, it isn't much more than annoying. Now let's shoot just one of these little balls out of a cannon at you. Ouch.

Photons are all moving at the same speed (the speed of light), but they contain different energies—what we've expressed before as frequencies or wavelengths. So a low-frequency photon doesn't deliver much kick, but a high-frequency photon can pack a wallop.

Unfortunately, there's a problem with this argument: It's relatively easy to prove that light absolutely, positively behaves like a wave, not a particle. The fact that light refracts (bends) when moving from one medium to another; the fact that it diffracts (like ripples in a pond interacting with a stick poking through the surface) . . . waves do these things, not particles, which travel in straight lines. But if

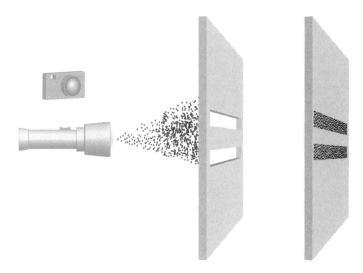

▲ Sending one photon toward a board with two slits and observing to see which slit it passes through, you can prove that it acts like a particle, traveling through one slit or the other.

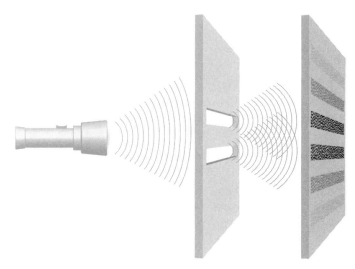

▲ If you don't watch, or if you send many through at the same time, the light acts like a wave, diffracting and creating interference patterns.

you fire light through two slits in a piece of cardboard, and use very careful measurements, you can determine that each photon is going through only one or the other slit, so there's also no doubt that it behaves like a particle.

You could dismiss this paradox, called the wave-particle duality, as just another wacky this-and-that fact of nature, like the old *Saturday Night Live* joke "It's a floor wax *and* a dessert topping." But actually, this is one of the greatest and most troubling mysteries in science today. It calls into question everything we think we know about the universe. We like to think that "stuff" is here or it's not, that it's matter or it's energy, but in fact everything is likely both: here and not, matter and energy. And if that doesn't confuse you, you don't understand it.

Fortunately, while the scientists and philosophers are arguing over the nature of reality, we needn't fully understand light in order to see it, measure it, and even use it to our advantage.

What We See The majority of what we humans understand is due to light, whether the reflections of the physical world around us or the glimmer of far-off stars. We gather information from what we see with our eyes and—perhaps even more important—our photosensitive instruments that can detect the invisible light around us.

Of course, color doesn't actually exist in the universe. We see color only because our eyes can register light at certain wavelengths and our brain attempts to make sense of those wavelengths by assigning them a visual meaning.

As we've seen, our ability to sense electromagnetic radiation is limited to a tiny range of wavelengths starting at about 380 nm (which we see as violet) and extending to about 750 nm (red). In musical notation, an octave is a doubling of a sound wave's frequency, and if you do the math, our spectrum of visible light equates to only about a single octave. Compare that to our 10-octave

"Yet mystery and reality emerge from the same source. This source is called darkness. Darkness born from darkness. The beginning of all understanding."

—Lao-tzu, *Tao Te Ching*

range of hearing, or the 45 octaves between AM radio waves and gamma rays.

Nevertheless, we can sense that little segment called visible light because of four particular types of nerve cells that we've developed in the tissue along the rear wall of our eyes: three types of cone cells and one rod cell—each named for its general physical appearance under a microscope. All are sensitive to light—that is, they can absorb electromagnetic energy and transmit it as a signal to the brain—but each is tuned to different wavelengths.

The three kinds of cones are most sensitive to light waves in the red, green, and blue frequencies, though each cone can also pick up a wide range of light. So, for example, the "green" cone can pick up some blue, yellow, and red, but it's most sensitive to the wavelengths we see as lime green. When light enters the eye, the cones quickly react, sending signals to the brain, which combines them—first finding edges (areas of widely different color contrast), then filling in the rest with color details until we determine what we're looking at.

Rods are far more sensitive, but they work best in very low light. Able to respond to even a single photon entering the eye, rods excel at night vision and for sensing very quick, small motions. They also tend to "wake up" more slowly than cones. For example, when you walk into a dark room, like a cinema, it can take several minutes for your eyes to adjust: Your cones aren't receiving enough light to function well, and your rods need time to get activated. After five minutes, your rods are working great, but you can barely make out any colors—just areas of light and dark.

Even more telling is the placement of rods and cones on the retina: Your eye contains about 6 million cone cells densely packed into the center, directly behind your pupil and lens. Surrounding the cones are about 100 million rod cells, like a ring around a bull's-eye target. This explains why, when walking outside on a dark night, you might see a star shimmer in your peripheral vision: The very dim

The only animal that can see both infrared and ultraviolet light is the goldfish.

"**The principal person in a picture is light.**"
—Edouard Manet

light hitting the outside edges of that target excites the rods but is often nowhere near bright enough to see when you try to focus your less-sensitive cones on it. Conversely, when you're in good light and you want to discern color or detail (such as these words), you need to look directly at it, focusing the image on your cones.

Granted, when we talk about humans, there are always exceptions. Some people lack one type of cone, resulting in what we call color blindness. They can still see color but must make do with only two signals instead of three. Conversely, some people—primarily women—have developed a mutant fourth cone. Where most of us are trichromats, these people are tetrachromats, able to see more colors—or, more accurately, more distinctions among colors, especially in the red to yellow tones. A tetrachromatic mother might be better suited to seeing tiny changes in a child's complexion or perhaps even notice subtle infrared heat radiating from a fever in a way that most of the rest of us could not.

It's unclear how many bands of color a tetrachromat sees looking at a rainbow. When Isaac Newton first used a prism to split white light into a rainbow in 1672, he named the five colors that most of us identify: red, yellow, green, blue, and violet. Later, however, in an effort to synchronize the spectrum with the seven notes of a Western musical scale, he somewhat arbitrarily inserted two additional bands of color: orange and indigo. Thus the primary-school mnemonic was born: ROY G. BIV.

Today, few people identify indigo as a hue separate from blue or violet. That doesn't mean we no longer see that color, but most of us probably don't call it out as notably different from the colors around it.

Curiously, we can identify colors that are not in the spectrum at all. Most notable is magenta—that hot pink color found on fuchsia flowers and in nearly every color printer in the world. You can easily create magenta on a computer screen by mixing red and blue light. Our eyes pick up the red and blue wavelengths and our brains mix them together. The result should be halfway between red and

> "The French philosopher Auguste Comte demonstrated that it would always be impossible for the human mind to discover the chemical constitution of the stars. Yet, not long after this statement was made the spectroscope was applied to the light of the stars, and we now know more about their chemical constitution, including those of the distant nebulae, than we know about the contents of our medicine chest."
>
> —Edward Kasner and James Newman, *Mathematics and the Imagination*

blue, but on the light spectrum that color is green! We can tell that the color we're seeing isn't green, so, in a that-does-not-compute moment, the brain makes up a color to see: magenta.

Messages in the Light If light is energy in motion, then it is information in motion, too. At its simplest, someone might light a bonfire on a hill fifty miles away in order to warn his tribe of danger—the information from the light is able to travel far faster than a messenger or even sound. If the tribe had particularly clever gadgetry, such as telescopes and prisms, they might even be able to learn what the folks on the hill were burning. This is due to a curious (but incredibly helpful) phenomenon: Different elements, when heated, give off specific wavelengths of light. Sodium gives off a different pattern of frequencies than carbon or hydrogen.

Armed with this knowledge, we can point our telescopes toward the sun and stars, carefully analyzing the light we capture and learning things we would not otherwise know: what the sun is made of, where black holes are hiding, how light follows the warped fabric of space-time as gravity bends reality. Our ability to tease apart light, to reveal its makeup, is called spectroscopy.

Of course, the sun and stars (and everything else in the universe) exhibit more than visible light. Radio waves and microwaves reach out across the cosmos, helping us map the solar system and the constellations. X-rays and gamma rays help us determine massive centers of otherwise invisible energy in the universe, such as pulsars and quasars. Everywhere we turn our instruments, we gather information, looking for understanding, searching for meaning.

And if we can find answers in the spectrums of natural light, then we can also encode our own messages in light that we create. The trick to doing this is called modulation: taking a known wave and adjusting it over time.

For example, let's say you tune in to an AM station at 700 on your radio dial—that's 700 kHz, or an electromagnetic wave oscillating back and forth at 700,000 cycles each second. AM stands

> **"What is essential is invisible to the eye."**
> Antoine de Saint-Exupéry,
> *The Little Prince*

Extremely low frequency (ELF) transmissions for naval communication frequency
3 Hz–3 MHz

Power lines
50–60 Hz

AM radio
520–1,620 kHz (1.62 MHz)

Shortwave radio
5.9–26.1 MHz

Garage door opener
40 MHz

Baby monitor
49 MHz

Radio-controlled airplane
72 MHz

Television stations
54–88 and 174–220 MHz

FM radio
88–108 MHz

Wildlife tracking collar
220 MHz

Cell phone
824–849 MHz

Cordless telephone
900 MHz

Global Positioning System (GPS)
1.2–1.6 GHz

Microwave oven
2.45 GHz

Short-range ("X band") radar tracking
8–12 GHz

Red light
400–484 THz (620–750 nm)

Yellow light
508–526 THz (570–590 nm)

Green light
526–606 THz (495–570 nm)

Blue light
606–668 THz (450–495 nm)

Violet light
668–789 THz (380–450 nm)

▲ Examples of light energy

for "amplitude modulation," which means the signal—the music, or the sportscaster yammering away—is encoded by increasing and decreasing the volume or intensity very slightly but very quickly. As we learned earlier, a light's intensity (its strength, or amplitude) is based on how many photons are being transmitted. Your radio senses those tiny changes and converts them into (you hope) a pleasant sound.

Switch to 103.7 on the FM dial, and your radio begins to pick up light waves at 103.7 MHz, or 103,700,000 cycles per second. Here you experience a different kind of signal: "frequency modulation," in which the frequency of the wave is altered up and down while maintaining its intensity. (More precisely, one wave with a varying frequency is merged with another, steady wave, and the result—an extremely complex wave—is transmitted, received, and pulled apart again.)

Awash in Light Radio broadcasts, satellite television transmissions, cell phone conversations, GPS signals, police and fire alerts, WiFi computer networks, radio-controlled toys, airport radar, garage door openers—we are constantly bombarded by light of varying wavelengths, energies, frequencies. And yet, strangely, most of this human-made radiation passes right through our bodies, even through the walls of our buildings, without affecting us or even slowing down. Why?

Light—which, remember, is made of photons, those tiny packages of energy that behave like both waves and particles—is unique in the universe in its ability to be both exceedingly tiny and extremely large. This range of size largely explains its ability to travel. Because each wave from an AM radio station is longer than a football field, the relatively small and sparse molecules that make up a wall don't have enough presence to impact it much. But line the wall with thick metal—a material with dense connections of electrically bound atoms—and those long waves can't get through.

You see, matter (like an atom or a molecule) can interact with a light wave in one of three ways: let it pass by, absorb it, or bend it. Which of these occurs is based on the wavelength of the light versus the type, size, density, and structure of the material. An X-ray has a tiny wavelength, so it passes through our skin like a bike zooming through a forest of widely spaced trees. Every so often it might hit a branch and cause a little damage, but in general it won't stop until it reaches a thick bramble, like the molecules in your bones. At that point, the light will likely be either bent (forcing a quick change of course) or absorbed (like a bike crash).

Microwaves, as we saw earlier, are far longer than X-rays—small enough to be easily absorbed by some food molecules but large enough that they cannot escape from the oven into your kitchen.

Visible light, on the other hand, consists of wavelengths that are just the right size to be absorbed or reflected by most matter around us. This is, as scientists like to say, evolutionarily advantageous: If our eyes were tuned to see radio waves instead, we'd be constantly banging into "invisible" things around us. But because we see the visible light spectrum, a ripe banana tends to absorb the "blue" wavelengths and reflect "red" and "green" waves into our eyes, causing us to see a yellow object.

Rocks appear solid because they, too, reflect and absorb various frequencies of light. But if you grind rocks to sand and melt the sand to make glass, you've changed the molecular structure so radically that wavelengths of visible light can now pass through it largely unimpeded. Yet those same molecules that let visible light pass through are also absorbing the slightly shorter wavelengths of ultraviolet light. So the size of the wave matters, but not as much as the material composition.

Now You See It . . . Light is such an inherent part of our everyday experience—whether through the colors we see or the heat we feel—that it's easy to forget how fundamental it is to the underlying structure of our universe, and how little we truly understand it. After all, electromagnetism is not just the study of magnets and generators and light waves. It is considered one of the four fundamental forces of nature, along with gravity and the strong and weak nuclear interactions. These are the most basic physical forces that hold our universe together and that cannot be described by any other, further reduced explanation.

Electromagnetic force pulls atoms into molecules and holds them together across enormous (to an atom) distances, forming what (from our size) appear to be solid objects. Without that subtle attraction, your chair, your floor, you, Earth would come apart as nothing but gas. At the core of this electromagnetic force is the lowly photon—called a quanta by Einstein as he laid the groundwork for quantum mechanics. Photons—known as the carrier particles of electromagnetism—are literally what make our universe possible.

Light—the oscillation of electromagnetic waves, the transmission of photons—is like the lifeblood of the cosmos, carrying packets of energy from one atom to another, or from one galaxy to another, through billions of years of space. Although it appears from our perspective that there is no medium in the vacuum between the stars (or between the atoms) through which these waves could be held and passed, that misses the point: Space itself is the medium. We are the medium. And we are the receivers.

SOUND

These go to eleven.

Nigel Tufnel, *This Is Spinal Tap*

SOUND IS MOVEMENT. SPECIFICALLY, SOUND IS THE MOVEMENT of molecules in a medium. And our sense of hearing is actually an extension of feeling—as though our ears were fingers that could reach out and stroke the ripples that run through the air around us. Hearing sound, like perceiving light, is another one of our evolutionarily clever ways of stretching our awareness beyond our boundaries, whether in search of food, concern for danger, or desire for connection.

If you stand next to a speaker playing loud music, you can often feel the sounds against your skin—rumbling low bass notes and buzzing high-pitched tones. You're feeling waves of air pressure, molecules moving and bouncing into each other. Our ears are so sensitive to these changes in pressure that we can detect when air molecules move one-tenth their diameter—that's about one-millionth the size of the smallest dust speck you can see.

The molecules move because of energy. Somewhere, somehow, something releases energy—a string is plucked, a dog barks, a reed flutters, or what have you. This is physical, kinetic energy; it moves, and the movement flows outward in waves. One molecule pushes into another, but the molecules can't move far until they affect another molecule, like people crowding to get out of a theater after the show.

So as one molecule pushes the next, each moves only a little, but the energy continues. It propagates. It travels. It transmits, like a

> **"I don't care much about music. What I like is sounds."**
> —Dizzy Gillespie, jazz musician

message being passed from one person to another. Each passing wears down the energy a little, but with enough thrust, enough power, it can travel a long way—perhaps even all the way to your ear.

The Medium Is the Message Of course, if there is no medium, there is no sound. The great Greek thinker Aristotle, student of Plato and teacher of Alexander the Great, first pointed out that we hear sound through the medium of air. Want the ultimate in soundproofing for your office or apartment? Encase yourself within a perfect vacuum of space, for where there is nothing (literally no-thing, no molecules to push around), there is no sound.

The statement in the movie *Alien* that "in space, no one can hear you scream" is true; those movie scenes in which inhabitants of one spaceship can hear a sound in another nearby spaceship are impossible. Without some medium—some solid, gas, or liquid to transmit the energy waves from one location to another—there is no message.

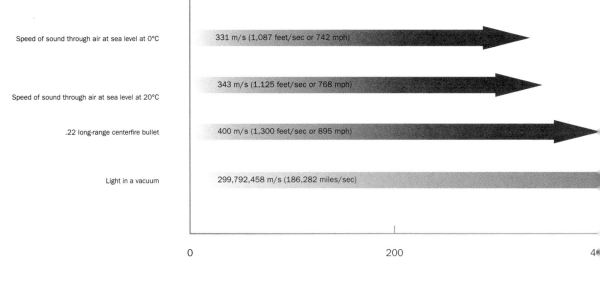

Speed of sound through air at sea level at 0°C — 331 m/s (1,087 feet/sec or 742 mph)

Speed of sound through air at sea level at 20°C — 343 m/s (1,125 feet/sec or 768 mph)

.22 long-range centerfire bullet — 400 m/s (1,300 feet/sec or 895 mph)

Light in a vacuum — 299,792,458 m/s (186,282 miles/sec)

0   200   4

This is distinctly different from electromagnetic energy. Electromagnetic waves—whether X-rays, radio waves, or even the small sliver of the spectrum called visible light—don't require molecules and can voyage through space indefinitely, easily traveling the 24 trillion miles (40 trillion km) of nothingness between us and our nearest star, Proxima Centauri. But a sound, no matter how great, peters out at the edge of the atmosphere.

Sound isn't limited to earth's atmosphere, of course. We are awash in the radiation of our sun, but we rarely consider the extraordinary noise that this ball of gas is making. You think a crackling fire is loud on a rainy night? The turbulence of gas and plasma churning, burning, exploding, at 16,000,000°C is staggering, sending incomprehensible shockwaves outward in all directions through the sun's mixture of hydrogen, helium, oxygen, and gaseous metals. The sound energy pounds out toward the surface of the burning ball, and then, as it reaches the boundary of the sun's atmosphere, the gas eventually stops, and with it the sound. Between our beneficent star and earth is silence.

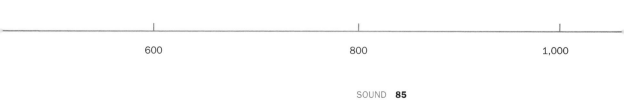

600	800	1,000

Of course, if you were actually close enough to hear the tolling of the sun, you would be vaporized. But just as you can "see" sound by watching a rattling window, scientists can see the sounds of the sun from 93 million miles away using precise Doppler measurements taken by the Solar and Heliospheric Observatory (SOHO) satellite. The surface of the sun vibrates with a complex set of resonances. Although the vibrations are too low for the human ear to hear, they can be sped up—compressing forty days worth of sound into a few seconds. The result is like an eerie bell, or a Buddhist bowl gong slowly, endlessly ringing into space.

Sound travels through water at different speeds, depending on temperature. So scientists can use hydrophones (underwater microphones) to determine water temperature by measuring the speed of sound in a particular location.

The Speed of Sound Because sound relies on molecules bouncing into one another, passing along their energy wave, it takes time for it to travel from one place to another. Sound moves fast, but not nearly as quickly as light—in fact, not even as fast as a bullet from a high-powered rifle.

It's relatively easy to measure the speed of sound: Place two people a mile apart, giving one a sports starter pistol with blanks and the other person binoculars and a stopwatch. When the pistol is fired (watch for smoke), start the stopwatch, then stop it when you hear the sound. Intuition says we probably couldn't start and stop it quickly enough, but you'll actually count almost five seconds before hearing the blast.

This exercise explains a common game during thunderstorms: Start counting seconds when you see a flash of lightning and stop when you hear the thunder. Divide the total number of seconds by 5 to find out how many miles away the lightning struck. (It's not 1 mile per second, as some people mistakenly believe.) Or divide by 3 to find the number of kilometers.

With careful measurements, you'll find that sound travels through air about 1,125 feet (343 m) in a second. That works out to about 67,500 feet (20,580 m) per minute, or 767 miles per hour (1,234 km/h).

Notice the word *about*. The speed of sound can change significantly depending on factors such as temperature. On a freezing cold day, it drops to 330 m/sec. On a hot day it can increase to 350 m/sec. The difference is due to the rate the molecules are moving through the gas we call air. As the sun warms the air, the molecules move faster, bounce into each other more often, and allow signals (such as sound wave energy) to transmit faster.

We need to be clear here: The molecules are not traveling very far. When a book falls off a table across the room, the affected molecules don't travel from the book to your ear. That would be wind, not sound. But like a series of billiard balls ricocheting across a pool table, the energy within the sound finds its way to you.

The makeup of the air also affects the speed of sound. For example, the speed of sound is faster in helium gas, which contains far lighter molecules than air, leading to the time-honored party game of talking with a lungful of helium. The pitch of your voice actually stays the same, but it moves through the gas faster, causing it to sound sped-up, like Donald Duck. If you're lucky enough to have a supply of heavier-than-air xenon gas handy, you can reverse the effect and sound like a "slow-talking cowboy" through the magic of slowing the speed of sound.

Sound can travel through liquids and solids, too, and at a very different speed than gas. The exact speed changes radically, based partly on the elasticity and density of the material. In freshwater, where the molecules are packed together in a slightly sticky and viscous solution, sound waves travel more than four times faster than in air: 1,482 meters per second (about 3,315 mph). In seawater, the speed of sound increases by a few percent, depending on temperature, depth, and salinity.

In a solid, where molecules are bound even more securely and one can barely move without affecting its neighbor, energy can pass even faster. A soft, relatively elastic material such as lead transmits sound waves at just over 2,000 m/sec, but in steel the speed of sound is about 5,960 m/s (that's 21,450 km/h or 13,300 mph). That's why you

Carbon dioxide	259
Air, 0°C	331
Nitrogen	334
Air, 20°C	343
Cork	500
Helium	965
Ethyl alcohol	1207
Water, distilled	1497
Water, sea	1531
Soft tissues	1540
Rubber	1600
Lead	2160
Gold	3240
Brick	3650
Marble	3810
Wood, oak	3850
Wood, maple	4110
Copper	4760
Glass, pyrex	5640
Steel, stainless	5790
Granite	5950
Aluminum	6420
Beryllium	12,890

▲ **Speed of sound in different materials (m/sec)**

can hear a train approach by pressing your ear to the tracks; you'll hear it 17 times faster through metal than through the air. If the tracks were made of an incredibly hard substance, such as beryllium or diamond, you'd hear it almost 40 times faster than through air.

Curiously, sound does not tend to pass very well from one medium to another. Sound waves act like light waves in this respect. Just as light reflects and refracts when moving from one medium (like air) into another (say, water), sound waves bend and bounce at these boundaries. So a crack of a hammer against the end of a long steel bar may travel beautifully through the metal, but little of it will transfer into the air on the other side. Similarly, sound waves from the air don't penetrate well into water, and vice versa. Anglers, take note: Feel free to chat with your buddies; the sound won't scare the fish!

A noisy object creates sound waves that expand out in all directions, like tiny ripples in a still pond. If the object starts to move, then those ripples keep expanding, but they become elongated, falling farther behind the object than in front of it, similar to the wake behind a boat. Just as waves don't hit the shore until long after the boat has passed, we don't hear the rumble of a jet airliner flying far overhead until the plane is almost out of sight.

But a funny thing happens as a jet plane nears the speed of sound: The sound waves compress and bunch up like wrinkled cloth in front of the nose of the aircraft. It's as though the molecules of air can't get out of the way quickly enough, and the pilot begins to experience increased aerodynamic drag, like an invisible hand pushing back.

When fighter jets first encountered this during World War Two, some erroneously believed that supersonic flight—flying beyond the speed of sound—was impossible, like traveling faster than the speed of light. However, the soldiers need only to have looked at their artillery. Bullets easily pass the speed of sound, flying through the air at speeds as fast as 5,000 feet/second (1,500 m/sec).

The exploding charge that propels ammunition creates a sound, but the bang actually stems from something unexpected: The

> "Wherever we are, what we hear is mostly noise. When we ignore it, it disturbs us. When we listen to it, we find it fascinating."
>
> —John Cage, composer

▲ Lockheed SR-71 Blackbird

pressure waves that build up in front of a supersonic object can no longer get away from their source. The result of this compression is an intense shock wave that travels through the air, often referred to as a sonic boom. Even a gun with a muzzle silencer cannot stop the sound of the bullet as it pierces the air, generating an intensely sharp peak in air pressure, like a thin but powerful wall of sound.

Modern fighter jets create similar, though obviously much larger, sonic booms as they move at transonic speeds. Though you may hear a sudden rumble that hits and then passes, the shock wave actually continues, following behind the plane like a sharp-edged sonic shadow that extends until the air pressure can sufficiently dissipate.

When dealing with very fast-moving objects, it's often helpful to discuss their speed as multiples of Mach, named in honor of the

A bullwhip snapped properly creates a tiny sonic boom as the end of the whip (the "cracker") quickly flips around faster than the speed of sound.

▲ Heinrich Hertz

physicist and philosopher Ernst Mach (1838–1916), who studied (among many other things) sound and ballistics. One half Mach is half the speed of sound, Mach 2 is twice the speed of sound, and so on. For many years, the fastest airplane was the Lockheed SR-71 Blackbird, which broke the speed record at Mach 3 in 1964. Forty years later, NASA's X-43A scramjet busted the doors off the record, with speeds up to Mach 9.6—nearly 7,000 mph.*

In order to escape Earth's orbit, a rocket needs to go even faster—about Mach 23 (17,000 mph), or twice that to propel itself to the moon. Obviously, the sound generated by these engines creates a sonic boom that is very, very loud.

How Loud Is Loud? Why are some sounds louder than others? Like ocean waves, sound waves are each created with a crest and a trough, and their sizes—the difference between their peaks and the surrounding air pressure (or sea level, using that analogy)—is called the wave's amplitude. For example, musicians know you take a small sound signal and make it huge with an amp, or amplifier—a device that increases the amplitude of the wave.

The bigger the difference in air pressure, the bigger the wave, the louder the sound.

We're not talking about huge differences in pressure here. Pressure is often measured in pascals (Pa), and we live in a bubble of air pressurized at 101,325 Pa. Let's say a sound wave is moving toward us. The air pressure momentarily increases and decreases by a tiny amount (remember, there is always a crest and a trough, technically called compression and rarefaction). If the pressure changes by 2 Pa (just 0.002 percent), we hear it—not as a whisper, as you might expect, but as the deafening sound of a jackhammer breaking through stone. A quiet conversation alters air pressure by as little as 0.0005 Pa (less than 5 ten-millionths of 1 percent).

*Though of course the X-43A and other recent hypersonic aircraft have all been unmanned, so the Blackbird could still technically be considered the fastest airplane.

Intensity of Selected Sounds

Loudness/Intensity (dB)	Source
<0	Silence
0–10	Faintest noise humans can hear
10–20	Normal breathing in quiet room, rustling leaves
20–30	Whispering at 5 feet
30–40	Library or calm room
40–50	Quiet office, normal talking or residential area
50–60	Dishwasher, electric toothbrush, rainfall, sewing machine
60–70	Air conditioner, automobile interior, background music, normal conversation, TV, vacuum cleaner
70–80	Coffee grinder, freeway traffic, garbage disposal, hair dryer
80–90*	Blender, doorbell, food processor, lawn mower, machine tools, noisy restaurant, whistling kettle
90–100	Shouted conversation, tractor, truck
100–110	Boom box, factory machinery, motorcycle, school dance, snowblower, snowmobile, subway train
110–120	Ambulance siren, car horn, chain saw, disco, jet plane on ramp, rock concert, shouting in ear
120–130	Heavy machinery, pneumatic drills, stock car races, thunder (short term hearing damage)
130–140	Air raid siren, jackhammer (threshold of pain)
140–150	Jet airplane taking off
150–160	Artillery fire at 500 feet
160–170	Fireworks, handgun, rifle
170–180	Shotgun, stun grenade
180–190	Rocket launch, volcanic eruption
194	Theoretical limit for undistorted sound in air

*Employers in the United States must provide hearing protectors to all workers exposed to continuous noise levels of 85 dB or above.

Perhaps you can begin to see how sensitive our ears are. Due to a complex system of interrelated bones and membranes in our middle and inner ear, we can pick up a change of less than a billionth of the atmospheric pressure, where air molecules are moved less than the diameter of an atom.

Instead of talking in pascals, most people refer to loudness using a different value: the decibel (dB), measuring one-tenth of a "bel" (named in honor of the telecommunications pioneer Alexander Graham Bell). The decibel is an almost entirely humancentric measurement: Zero decibels marks not some universal constant but the lower limit of human hearing—the faintest sound we can detect. Below that, you can't tell the difference between sound and air molecules just randomly bumping up against the eardrum.

The logarithmic nature of the decibel system means that for each additional 10 decibels, *ten times* more power is required, but it *doubles* the perceived loudness of a sound. That is, a normal conversation (about 40 dB) is about twice as loud as a quiet library (about 30 dB), but that 30 decibels reflect 1,000 times more power than near silence. A large truck driving by can throw 94 decibels—carrying almost ten million (10^7) times the power of a whisper.

In a famous 1976 concert, The Who was measured at 126 dB, 100 feet (30 m) from the stage. More recently, the band KISS hit 136 dB—the equivalent of standing next to a jet airplane taking off—during a 2009 Canadian concert, just before being forced to "turn it down" by local law enforcement. That's 17,000 times louder and 10 trillion times more powerful than a heartbeat. One can only hope earplugs were liberally distributed before these shows, as permanent hearing damage can be caused by sound above 120 decibels.

Of course, a rock concert is nothing compared with the new international "sport" of dB drag racing—where competitors build cars that contain virtually nothing but an engine and audio equipment. The goal is to create the loudest car, if only for a few seconds. The vehicles have two-inch-thick windows and doors bolted and clamped closed so as not to rattle off their hinges. Participants

> "You can tell a good putt by the noise it makes."
>
> —Bobby Locke,
> South African golfer

stand outside and throw a switch to create a pulse of noise so powerful it can literally melt the metal in the speakers. The world-record car, at about 180 dB, is 60 times louder and reflects a million times more power than a typical concert.

In fact, that car nearly ties the loudest sound on record, which most historians identify as the 1883 volcanic eruption at Krakatau, Indonesia. The cataclysm, in which most of the island was destroyed and ash was propelled 50 miles (80 km) high, had an intensity of just over 180 dB and was audible 3,000 miles (5,000 km) away in Mauritius. The shock wave traveled even farther, reverberating literally around the world over the next five days.

But could a sound be even louder than that? It depends on how you define *sound*. There is no maximum strength of a shock wave. Blow up a few hundred pounds of TNT, and you'll create about 200 dB of pressure—a wave of such power that it would likely kill any human nearby. Nuclear explosions reach over 275 dB. However, if you limit the definition of sound to a pressure wave with a crest and a trough, a signal that conveys a message through the air beyond a distorted, deadly boom, then you're limited to a wave no bigger than atmospheric pressure itself. That is, the rarefaction (the low point on the wave) cannot drop below zero Pa, the pressure in a vacuum. And a drop from normal atmospheric pressure of 101,325 Pa to zero Pa results in a maximum possible sound volume of 194 dB.

Sound power is measured in watts per square meter or W/m². The ratio of power required to generate the faintest sound we can hear, up to the level of "Ow, that hurts!" is 1:100,000,000,000,000 (a hundred million million). This is yet another example of how incredibly dynamic and sensitive human hearing is. **The farther from a sound** source, the less loud you hear it. Specifically, the sound intensity drops by about 6 dB each time you double the distance.

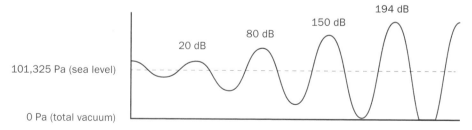

▲ As volume increases, sound waves eventually get clipped.

What's the Frequency, Kenneth? Our ears are clearly sensitive to a sound wave's amplitude, but they're just as sensitive to the distance from the crest of one wave to the crest of the next—the wavelength. This measurement, along with the speed the sound is traveling, determines the amount of time it takes for one full wave to crash into our eardrums, like waiting for one ocean wave to finish before the next comes ashore.

That math-loving Greek Pythagoras, about 2,500 years ago, first observed that a string held taut and plucked vibrates at a particular rate. The vibration of the string then causes a sound. But tighten the string, use a thinner filament, or shorten it, and the pitch rises to a higher note. A string half as long produces a sound exactly one octave higher. A string twice as long is an octave lower.

A string is a helpful tool to understand waves because we can literally see it quiver. A loose, flapping wire makes no sound we can hear, but tighten it until it's vibrating more than 20 times per second, and you perceive a moan; keep tightening, and the sound rises to a groan, then a growl, each slightly higher in pitch. The tone is based entirely on the frequency of the waves—that is, the number of vibrations each second. Lower frequencies—longer waves and fewer cycles per second—sound lower to us. Higher frequencies (shorter waves, more coming at us each second) sound high-pitched.

We see examples of vibrations all around us. A bird flapping its wings 2 or 3 times per second creates no sound we can hear. But a bumblebee wing, flapping about 200 times per second, creates a low hum. A mosquito wing, moving 600 times per second, is an annoying whine. Again, our hearing comes to our rescue, enabling us to "reach out" and find the insect invader.

Scientists replace the phrase "waves or cycles per second" with the simple word *hertz* (Hz). You could say a clock ticks off time at 1 hertz (once per second), though the tick or tock it makes is actually a sound wave vibrating far faster. In fact, the lowest pitch humans can hear is a wave vibrating around 15 Hz. A tone at 30 Hz sounds about the same, but an octave higher; the same can be said for 60, about

Sound Frequency

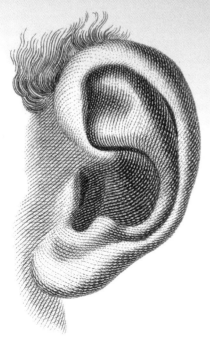

Frequency	Sound phenomenon
0.1–2 Thz	SASER (sound laser, in development)
1–20 MHz	Medical ultrasound
25–100 kHz	Bat sonar clicks
40–50 kHz	Ultrasonic cleaning
32.768 kHz	Quartz timing crystal
18–20 kHz	Upper limit of human hearing
4–5 kHz	Field cricket (Teleogryllus oceanicus)
2048 Hz	C7 scientific scale, highest note of a soprano singer (approximate)
440 Hz	A4 American standard pitch, TV test pattern tone
435 Hz	A4 international pitch
261.63 Hz	"Middle C" (C4 in the American standard pitch)
256 Hz	C4 scientific scale, typical fundamental frequency for female vocal cords
128 Hz	C3 scientific scale, typical fundamental frequency for male vocal cords
64 Hz	C2 scientific scale, lowest note of a bass singer (approximate)
50 Hz	Ruby-throated hummingbird in flight
20–50 Hz	Cat purr
20 Hz	Lower limit of human hearing
17–30 Hz	Blue and fin whales are the loudest marine sound in this range
1–5 Hz	Tornadoes

the throb of a hummingbird flying by. Double that is 120, about the typical frequency of a man's voice. Another doubling brings us to the range of a woman's voice, though the human voice can actually span from about 80 to 1,100 Hz.

When the sound waves are compressed into the thousands-per-second range, we start measuring in kilohertz (kHz). Children can easily hear sound up to 20 kHz (20,000 wave cycles per second), though that ability tends to wear out for one reason or another until, at middle age, we tend not to be able to hear anything higher than 15 or 16 kHz. Marketers have taken advantage of the difference. A Welsh security company created the Mosquito sound repellent that emits screeches in the 17 kHz range, designed to repel teenagers from loitering in front of shops. Of course, the tables were soon turned when the high-frequency sound was converted into a cell phone ringtone—one that kids can hear but adults (such as teachers and parents) cannot.

Note that this range, from 20 waves per second to 20,000, is an extraordinary span, reaching over ten octaves (where each octave

Weather conditions can also make sound travel farther or shorter distances—but not for the reason you think. A strong wind won't carry a sound wave along faster or push it backward like it would a material object. After all, the speed of sound is far faster than the speed of the wind! However, wind shear—when the direction that air is traveling changes suddenly from one part of the atmosphere to another—can affect how sound travels. Sound typically bends upward as it travels out, but with an appropriate wind shear, it will reflect downward again, like a stone skipping on a pond—causing someone farther away to be able to hear it. Similarly, temperature inversions—where a layer of warm air rests on top of a bubble of cooler air—can also cause loud sounds to travel much farther than expected. An explosion at an English oil depot could be heard a surprising 200 mi (320 km) away in the Netherlands as the sound ricocheted off a layer of air. People who were much closer to the blast did not hear it, as the sound traveled over their heads, creating a "sound shadow."

represents a doubling of frequency). Compare that with our eyes, sensitive to only a single octave of the electromagnetic spectrum between about 400 and 780 terahertz.

Plus, imagine the speed at which our ears are processing information. First, the sound waves are captured by our pinna (those are the fleshy things sticking out from the side of your head), which act as sophisticated sound-processing gear, cleverly amplifying and filtering sounds before focusing them into the ear canal. The waves then vibrate a thin but rigid piece of skin, technically called the tympanic membrane but commonly called the eardrum.

While the analogy of a drum seems apt at first, with the sound waves beating against it, the truth is that the compression and rarefaction (the positive and negative changes in pressure) actually push and pull at the eardrum. This physical movement is then transferred to an astonishingly complex mechanism in which the three tiniest bones in your body (you remember from school: hammer, anvil, stirrup) act like a hydraulic lever, amplifying the faint sound signals 22 times while pressing against the fluid-filled snail-shell-shaped cochlea. Finally, the waves pass through the cochlear fluid to stimulate more than 20,000 minuscule hairs, like underwater currents wagging long strands of seaweed attached to the ocean floor. A sound's wavelength translates directly into how far along the cochlear spiral the wave breaks, exciting the hairs. High-frequency sounds release their energy by moving hairs early on; lower-frequency waves stimulate hairs farther along.

Finally, the hairs convert their movement into electrical signals and send them on to the brain. And it all happens in an instant.

Echo, Echo When we explored light and its wavelengths, we were dealing with extremely small fluctuations in electromagnetic radiation—sizes on the order of millionths and billionths of a meter. Sound waves are far longer. Even the highest pitch we can hear—the one with the most vibrations per second and therefore the smallest waves—represents a wave about 1.7 cm long, from crest

> "There's one thing I hate! All the noise, noise, noise, noise!"
> —Dr. Seuss, *How the Grinch Stole Christmas*

> "The world is never quiet, even its silence eternally resounds with the same notes, in vibrations which escape our ears."
> —Albert Camus, *The Rebel*

> "There is geometry in the humming of the strings, there is music in the spacing of the spheres."
>
> —Pythagoras

to crest. That's about 20,000 times longer than the longest visible electromagnetic wave we can see, red light!

You can easily calculate a wavelength by dividing the speed of sound by the frequency of the sound. So the musical note middle C, with a vibrational frequency of about 262 Hz, corresponds to a wave about 4.3 feet (1.3 m) long. The lowest pitch humans can hear is an astounding 75 feet (22 m) long.

Of course, traveling at 343 m/sec, these long waves, with their alternating increase and decrease of air pressure, still roll over us in a matter of milliseconds.

The length of a sound wave affects how we hear it in another way, too. When a wave of any kind hits an obstacle, the result depends on the wavelength. A wave shorter than the obstacle tends to reflect off it. We can see that easily with light waves, far smaller than even the tiniest object we can see with our eyes. Shine a flashlight on something and it causes a shadow where the light does not reach.

The same thing happens with sound waves: You can shout close to a wall and hear its reflection. Submariners take advantage of this effect, navigating the murky depths with sonar—flashes of sound that echo back the location of large objects in the water.

But due to their size, sound waves don't reflect off small things—anything smaller than the size of the wave itself. Instead, they diffract—they bend around. That's why you can hear music from a stereo down the hall, though it tends to sound muted with too much bass. The higher-frequency short-wavelength sounds get blocked at doorways and don't diffract as well. Lower-frequency sounds—those with longer wavelengths—tend to bend around corners quite well, allowing them to travel far and wide.

This is also why you can close your eyes and turn your head left to right and you are able to locate where in the room someone is speaking: Your own head literally makes a sound shadow, and the higher frequencies of the voice go in one ear more than the other. Similarly, stereo enthusiasts know that it doesn't matter where in a room you put a low-frequency subwoofer; the huge wavelengths

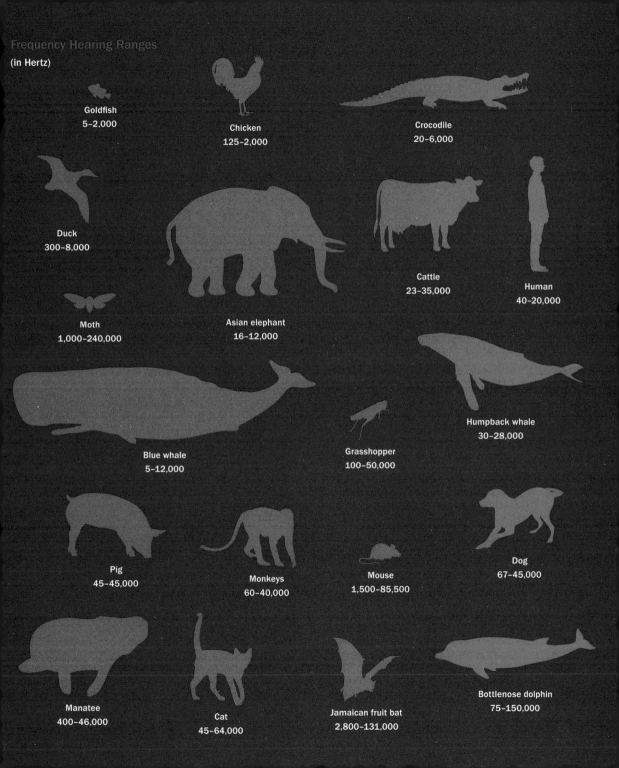

Frequency Hearing Ranges
(in Hertz)

Goldfish
5–2,000

Chicken
125–2,000

Crocodile
20–6,000

Duck
300–8,000

Moth
1,000–240,000

Asian elephant
16–12,000

Cattle
23–35,000

Human
40–20,000

Blue whale
5–12,000

Grasshopper
100–50,000

Humpback whale
30–28,000

Pig
45–45,000

Monkeys
60–40,000

Mouse
1,500–85,500

Dog
67–45,000

Manatee
400–46,000

Cat
45–64,000

Jamaican fruit bat
2,800–131,000

Bottlenose dolphin
75–150,000

diffract so much that you're unlikely to be able to tell if the source is in front of or behind you.

There's another reason sound quality changes through space: Matter (the air, your chair, whatever) absorbs sound energy, converting it into a hardly detectable amount of heat. A large room or a deep canyon may echo every sound, but the quality of the tone tends to be somewhat muted because higher-frequency smaller-wavelength sounds are absorbed more quickly than those low-frequency thumpers. This explains why a thunderclap sounds like a sharp crack when it's near you but only a low rumble a mile or two away.

Beyond Our Sound The fact that as we age we naturally lose the ability to hear higher frequencies may make you wonder if there are other sounds "out there" that you're not hearing. The answer is absolutely, though they're not necessarily sounds you want to hear.

Sound waves at frequencies higher than 20 kHz are called ultrasonic (that's different from supersonic, which means faster than sound). Dogs can hear ultrasonic vibrations up to 45,000 cycles per second, and cats probably hear a bit beyond that. The reason is likely evolutionary: If you're hunting a small animal like a mouse, you want to hear it—and the call of a young mouse in distress can easily hit 40 kHz.

Hunting by sound is also the key to a practice known as echolocation. If you can make a sound wave narrow enough to reflect off whatever it is that you want to eat, you can find it—sense it, almost feel it—by listening to the echoes around you. The classic example is bats, who can bark at well over 100 kHz. The sound is very loud and very short (usually only a few milliseconds in duration), but the wavelength is just right, bouncing off anything as small as 2 or 3 millimeters—quite helpful when looking for insects or avoiding the branches of trees. When it comes to larger animals or objects, bats can even "see" small features, letting them know what something looks like, what kind of animal they're approaching, and so on. Even more astonishing is the fact that bats often fly in packs of hundreds or even thousands but can still navigate by recognizing their own voices.

Humans have found a number of clever ways to make use of ultrasound. Dentists use it to clean teeth, doctors focus it to break up kidney stones noninvasively (a practice known by its tongue-twister name *lithotripsy*), therapists use it to apply "deep heat" muscle treatments, and diagnosticians and engineers use ultrasonography to visualize the interior of the human body as well as test the structural integrity of plastics, wood, or metal beams. However, these practices all use sound waves with frequencies far beyond those audible to animals, with waves reaching from 50,000 cycles per second up to 18 million hertz (18 MHz).

At the other end of the sound frequency spectrum, below 20 waves per second, sits the mysterious world of infrasound. Whereas dolphins, porpoises, and orca whales echolocate in the ultrasonic range, when it comes to communication, most marine mammals tend toward this lower end of the sound spectrum. As we've seen, lower-frequency sounds can travel farther, and those below 1,000 Hz happen to travel much farther in the saltwater sea. So the humpback and blue whale can sing out extremely loudly (over 150 dB) in the range of 10 to 30 Hz—low, powerful songs that can travel hundreds of miles.

On land, elephants, hippopotamuses, and alligators also use these infrasonic tones (sounds with a frequency too low for humans to hear) to communicate with their brethren, allowing for widespread coordination of herds, or for males to find mates. A female elephant, for example, signals its availability by creating distinctive rumbling noises that can broadcast over several kilometers. Zoologists report they can feel these calls thrumming through the air even when they cannot hear them. It's unclear how the animals themselves detect

Humpback whales sing songs so loudly that they can be heard more than 100 mi (160 km) away. However, it is sperm whales that emit the loudest sound of any animal, using an air-blowing structure in their heads, curiously named "monkey lips." These clicks can be as loud as 230 dB!

these noises, though as pressure waves travel better through rock than air, it's possible that they feel the sound through their feet.

Curiously, even though we humans cannot hear infrasonic tones, we can detect them, and the effects can be dramatic. In 2003, a team of researchers in London set up an "infrasonic cannon" in the back of a concert hall, adding very soft (only about 7 dB) and very low (17 Hz) sounds intermittently to several pieces of music played before a large audience. When asked, 22 percent of the listeners reported feelings of intense discomfort, fear, or—on the flip side—a sense of the supernatural or numinous, using phrases such as "an odd feeling in my stomach," "feeling very anxious," and a "strange blend of tranquility and unease."

The fact that infrasound makes humans uncomfortable has been repeatedly documented. Employees have refused to work in certain factory rooms in which they felt inexplicably ill, until it was discovered that vibrating cooling fans were pumping more than just air into the environment. Some scientists now believe that many reports of haunted houses actually stem from underlying and hard-to-trace infrasonic waves. For example, a sound wave at just the right frequency—about 18 Hz—can actually cause the human eye to vibrate, and this vibration may cause mysterious gray apparitions in the peripheral vision. Could ghosts be what Shakespeare's Macbeth called "full of sound and fury, signifying nothing"?

The lowest sounds ever found—far below the infrasonic thunder of animals, avalanches, or earthquakes, measured at even less than 1 hertz—are those of distant cosmic events. Many galaxies contain a huge quantity of "free-floating" gas—the residue of untold numbers of stars that have grown and exploded over billions of years. Astronomers, looking at a black hole in the Perseus cluster of galaxies, about 250 million light-years away, recently noticed a pattern in the clouds. More dense in some areas, less dense in others, the pattern soon revealed itself to be a sound wave, emanating from the black hole.

Many species have developed specialized organs to produce or detect sounds. Arthropods such as spiders and cockroaches have special hairs on their legs that can sense sound. The antennae of mosquitoes and many other insects can sense minute variations in air pressure. Honeybees appear to communicate through the buzzing vibrations of their wings; ants, crickets, and even some snakes and spiders stridulate (rub body parts together) to create chirps, clicks, or hisses. Some of these can be extremely loud: The African cicada's sound has been measured at over 105 dB!

The sound is a single note, drawn out not over meters but over billions of meters. To be precise, the researchers have determined that it's a B-flat 57 octaves below middle C, a million billion times lower than the lowest sound we can hear. If you can imagine, where a 20 Hz sound wave would take 1/20 of a second to pass by, a single wave of the Perseus black hole's drone would take 10 million years. Truly, the irony is palpable; Plato noted that "the empty vessel makes the loudest sound."

Of course, as we've seen, sound is ultimately absorbed and converted into heat. And scientists estimate that these tones, these rich, vast roars, provide as much energy throughout a galaxy as billions of suns. It's as though the music of the spheres, heating this interstellar gas, helps create just the right conditions for new stars and galaxies to be born.

Complex Sounds Today, on our small planet, we are saturated with an astonishingly rich sound field—one in which the spectrums of frequency and amplitude are woven together alongside rhythm, cadence, harmony, and many other audible ingredients. Even more amazing is that we can make sense of the melee. You may be in conversation with someone at a dinner party, overhearing another discussion at the table, tapping your toe to the music in the background, and suddenly catch the cry of a baby in the next room.

One reason we can distinguish among these various signals is that sound is rarely pure. If a flute and a piano each played a slightly flat A by emitting a perfect 436 Hz frequency, we'd never tell the difference between them. But musical instruments, and most objects that produce a sound, create overtones—combinations of additional frequencies, usually at even multiples above the fundamental, lowest note, called harmonics. So play a flute—essentially a metal tube with some holes in it—at 436 hertz and you'll invariably find a strong second tone at 872 hertz added to the mix, along with a dash of 1,308 Hz and 1,744 Hz. You'll also hear a faint jumble of many other frequencies between, creating the characteristic breathy sound. A

In the beginning there was silence, and then, suddenly, there was light. When the astronomer and science-fiction writer Fred Hoyle coined the term "big bang" in 1949, he didn't intend to describe the sound of creation, and in fact it's a misnomer—with no medium, the explosive expansion of the new universe would have been a light show without a sound track. But it wasn't long before vibration began to pulse through the blinding sea of photons, and then later the primordial soup of early atoms. The effects of these earliest waves can be detected even today, 13.7 billion years later, as astronomers map the heavens. We can see galaxies—each filled with billions of stars—clumping together every 500 million light-years or so in alternating crests and troughs, compression and rarefaction.

piano combines the same ingredients in very different amounts to concoct a completely different flavor. The exact blend of frequencies helps define what we call the instrument's tone, or timbre.

Nevertheless, the processing required to decipher the billions of overlapping waves we hear during a few moments at that dinner party seems beyond possibility. Yet it's just another day at the office for our ears.

And, it should be noted, for our skin—for even the deaf can sense and appreciate sound waves. Many deaf people enjoy dancing to the rhythms of loud, rumbling bass-laced music that they can feel. At music concerts, some deaf audience members hold balloons between their fingertips that act like external eardrums, resonating and amplifying the sound, allowing a richer appreciation.

While at first glance this may seem like a completely different activity from hearing, remember that hearing *is* feeling. In fact, brain researchers have recently discovered that deaf people experience these physical vibrations in the same part of the brain where hearing people process sound from the ears. It's clear that humans are designed to detect and understand sound, one way or another.

There is no doubt that sound is a crucial part of our human experience, and perhaps even beyond. Virtually every faith tradition focuses on the creative and healing power of sound. In the Sufi Muslim tradition, music and chant are the secret of bringing one closer to Sirr, the center of inner consciousness where contact with the Divine is possible. As the influential twelfth-century Islamic philosopher Abu Hamid al-Ghazzali said, "There is no way to the extracting of [the heart's] hidden things save by the flint and steel of listening to music and singing, and there is no entrance to the heart save by the antechamber of the ears."

Similarly, Christians and Jews focus on the "Word" of spiritual revelation, reflected through the prophets or—in the mystical traditions—our own chants, purportedly generating heavenly reverberations. Eastern religious practices include meditative intonations of mantras, often based on the "universal sound" of *aum* or *om.* This is an ancient idea—that sound can be both viscerally and metaphysically transformative, resonating within ourselves and to the celestial heights.

And the idea is certainly not without merit, as the phenomenon of resonance is easily shown. Every object, every material, has a natural frequency at which it vibrates. For example, tink the side of a wineglass to hear its special pitch. If you play a matching tone, your sound's waves add to the object's—make it loud enough and you can cause the material to shake to the point of breaking. The stories of singers shattering glass goblets are true!

You can see this trick of physics in any kind of wave. Pushing a child in a swing at intervals that match the swing's resonant frequency makes it go higher, even with very little effort. Push at a faster or slower rate, and neither you nor your child will enjoy it. A piano or cello string does the same thing all by itself: Play a similar note on another instrument—either the same pitch or one that shares the same harmonic overtone—and the string sings out in reply.

So who is to say if the sound of song or prayer could not excite that which is beyond our seeing?

As our ears capture signals from the waves, we become aware of our interconnectedness with the matter around us. A footstep, a word, our favorite song, a brush of silk—we discover significance in the sounds that travel to us and through us, just as we create sound to convey meaning and relationship. These cycles of energy are powerful influencers, from the dark rushings of the mother's womb to the explosive death of stars.

HEAT

It doesn't make a difference what temperature a room is, it's always room temperature.

—Steven Wright

THERE ARE FEW THINGS MORE PLEASURABLE, ON EARTH OR IN the heavens, as a good hot bath or shower. Its enveloping warmth is reassuring, replenishing, rejuvenating—and for a good reason: Heat implicitly means life. Life requires heat, though like the fable of Goldilocks's bears, not too much heat, not too little, but just the right amount. The heat from a bath or a lover's embrace is a reminder that life is good and, if only for a moment, all is well.

Heat, at its essence, is motion—the motion of atoms and molecules goaded into movement by electricity, compression, chemical reactions, nuclear forces, or one of many other sources of energy. Energy inevitably turns into motion, like kids on a sugar high, and the motion spreads from atom to atom, molecule to molecule, until the warmth is shared as equally as possible.

Temperature—the measure of heat—is our natural way of gauging how much energy is in something. An ice cube has little energy to offer. A sweltering hot day is buzzing with energy, though the ambient humidity may make you feel as though you're drained of yours. We are constantly aware of temperature, because every aspect of our life and health depends on it.

And yet we actually recognize only a small sliver of the wide range of temperatures in nature. Our fragile bodies can handle only the smallest variation in heat. An object only 30 degrees greater than our own internal temperature can cause significant burns, and if our own body temperature drops even 10 degrees for any significant amount of time, the result is catastrophic. Of course, our warm-blooded metabolism lets us regulate our body heat appropriately, so we perspire to cool ourselves, or generate cell heat as required, even resorting to the wild movements of involuntary shivering if necessary. But if these systems fail, either hyper- or hypothermia can set in, shutting down key chemical reactions in your organs and ultimately leading to death.

Other animals have adapted to deal with heat fluctuations in other ways. The North American wood frog doesn't even try to get warm when winter sets in. Instead, as the temperature drops, it suffuses its cells and bloodstream with a cocktail of sugars and proteins that allows it to freeze solid without tissue damage. Once frozen, it shows no signs of life whatsoever: no heartbeat, no breathing, no kidney function. It is as dead as a stone . . . until the spring thaw, when some deep unknown signal miraculously tells everything to start up again, and in a matter of hours the frog is hopping about looking for a mate.

We've learned to survive in the harshest of both arctic and desert conditions, but even these are temperate compared with some places in space, or deep inside the crust of our planet. Temperature is energy in motion, and energy—as you may guess—covers a wide gamut.

A bolt of lightning can reach 50,000°F—hotter than the surface of the sun—and packs a punch between 100 million and 1 billion volts.

Measuring Temperature If you were traveling to Sweden and heard the temperature was 22°, would you want to bring a jacket? Would you worry if, instead, the forecast read 295°? It all depends on the scale you're using, of course: Celsius, Kelvin, or Fahrenheit.

Scientists as early as the second century BCE discovered that certain substances, such as air, expand when heated, but it was not until the seventeenth century that scientists such as Galileo Galilei used this knowledge to build devices that would measure heat itself, called "thermo-meters." When Isaac Newton made the radical suggestion to place marks on the thermometer in order to better record specific values, he prescribed that the zero mark

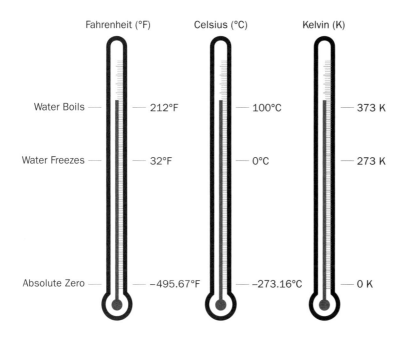

Fahrenheit (°F)

Celsius (°C)

Kelvin (K)

Water Boils — 212°F — 100°C — 373 K

Water Freezes — 32°F — 0°C — 273 K

Absolute Zero — −495.67°F — −273.16°C — 0 K

should indicate melting ice and 12 reflect the temperature of the human body.

The value of 12 may seem odd (why not 10 or some other reasonable number?), but note that this duodecimal system is particularly handy when it comes to splitting evenly into sixths, quarters, thirds, and halves. Thus, the English standardized on 12 inches to a foot, 12 pence to the shilling, 12 units in a dozen, 12 dozens in a gross, and so on.

In 1714, a young glassblower-cum-physicist named Daniel Gabriel Fahrenheit hit upon several genius improvements. Instead of using sticky, imprecise liquids such as alcohol inside the thermometer, he used mercury. In order to encompass a wider range of values, he set the zero mark at the melting point of saltwater, which freezes at a significantly lower temperature than freshwater. And he enabled finer increments by increasing the top value, the body's temperature, to 96°. (Once again, the choice of 96 makes sense only when you notice that it's easily divisible by 2, 3, 4, 6, 8, 12, and so on.)

A few years later, as scientists decided the boiling point of water was a more important value than body temperature, Fahrenheit's scale was fudged slightly. Water, it was declared, should boil at exactly 180 degrees above its freezing point of 32°, at 212°F. This adjustment was convenient (180 is also easily split into smaller fractions and matches up nicely with the number of degrees in a half circle), but this required stretching Fahrenheit's scale a little bit, so that our body temperature would now be marked at the awkward value of 98.6°F.

Given that the melting and boiling points of water are such important measurements (at least here on Earth), why not set those values to 0 and 100? Such was the reasoning of the Swedish astronomer Anders Celsius in 1742. (Actually, to be accurate, he bizarrely set freezing at 100 and boiling at 0, but that was quickly rectified a couple of years later, soon after he died.) Because each division on this thermometer measured exactly one hundredth of the total scale, it was labeled as a "degree centigrade" (Latin for "hundred steps"). That nomenclature stuck for three hundred years, until, in 1948, the term was changed by the International Committee for Weights and Measures to "degrees Celsius."

Now that we had two different ways of measuring heat, why bother with a third? By the mid-eighteenth century, scientists had realized that there was a world far beyond that of the boiling and freezing of water. At first, there didn't appear to be any limit to how hot or cold a substance could get. After all, if you could heat something to 1,000° Celsius (the temperature in a typical stovetop flame), then why not cool it to −1,000°C?

Unfortunately for this theory, as people were finding clever ways to cool nitrogen and other gases to their freezing point, they discovered a curiosity: For each degree Celsius you cool a gas, it reduces in volume a tiny amount—about 1/273. That led them to an

▲ Anders Celsius

To convert Fahrenheit to Celsius, subtract 32, multiply by 5, and then divide by 9. To convert °C to °F, multiply by 9, divide by 5, and then add 32.

intriguing conclusion: At that rate, the gas would disappear entirely, or at least take up zero space, when it reached –273°C. While their understanding of the science was still immature, they did correctly surmise that this must signify the coldest temperature possible.

The data was compelling enough that the Scottish physicist William Thomson—who had gained fame and a knighthood for his work on the telegraph—suggested that this lowest low should be the new zero, an *absolute* zero. Sir William's idea stuck, but you never hear about "degrees Thomson." Rather, he was later raised to the House of Lords, assuming the heady title Lord Kelvin of Largs. (He lived in Largs, and his office was on the river Kelvin, which flows through Glasgow.) So scientists began to talk in terms of "degrees Kelvin"—a system in which each degree is the same "size" as a degree Celsius, but in which zero starts much lower.

Because heat is motion, and motion is energy, and there are many different forms of energy and reasons to discuss them, science has developed a number of other ways to describe and discuss heat in a system. We talk about calories, for example, and when discussing the heat from a burning gas we might talk about joules or numbers of BTU (British thermal units), where a single BTU describes the energy given off by a wooden kitchen match.

Fortunately, we don't need this alphabet soup of heat descriptors to talk about how hot a proper cup of tea should be, or the weather on Venus. For common, daily usage, degrees Fahrenheit or Celsius do fine, and even when exploring the very cold or very hot, all but the geekiest discussions can use kelvins.

The Chaos Meter Here's an amazing magic trick you can try at home: Place a chunk of ice on a plate and leave it out on the counter. By mumbling numerous incantations (and just waiting awhile) that solid object transforms into—gasp!—a liquid. Wait long enough and that clear puddle—double gasp!—appears to vanish completely. Like any good magic trick, it seems miraculous until you know how it's done,

▲ Sir William Thomson, later Lord Kelvin

Scientists attending the 1967 General Conference on Weights and Measures gave Lord Kelvin one of the greatest honors of all when they agreed to drop the word *degree* and just call the measurements "kelvins." Thus, he joined a select group of inventors whose names have become uncapitalized measurements, such as watt, volt, ampere, and joule.

at which point it becomes ordinary. But allow yourself a moment to see the magic through fresh eyes. A solid turning liquid turning gas, a seemingly insignificant element: heat.

For centuries, scientists assumed that heat was literally an element—an invisible fluid called "caloric" that traveled from object to object. The seventeenth-century English philosopher and father of liberalism John Locke suggested that heat was a form of kinetic energy—the motion of the tiny "insensible parts" of a substance.

If heat is motion (which we now know it is), then there is technically no such thing as "cold." Obviously, one object may feel colder than another, but what you're really talking about is heat—there is only heat, which can be added and removed, motion sped up or slowed down. In other words, one object isn't really colder than the other; it's just less warm.

And here's the clever part: The measure of heat is also the measure of how much chaos or order there is. Cooling a substance removes energy, allowing molecules to rest into rigid structures. Adding heat smashes up that architecture, leaving a somewhat dense broth; adding even more heat allows the molecules to break free from one another, a rapidly expanding gas of entropy like a flock of birds rushing into the sky. Even the word *gas* itself stems from a Dutch pronunciation of the Greek word *chaos*.

It's important to remember that at the atomic level, nothing ever stops moving. On a pleasant spring day, air molecules are flying at 1,150 mph (1,850 km/h), manically bouncing off one another's magnetic fields, gently buzzing your skin, generating pressure and transporting heat. Even in a solid, like a crystal in which each atom is held tightly in place, molecules never stop dancing, wriggling atom to atom along their internal degrees of freedom. And, in turn, each atom vibrates with activity, each electron a never-ending now-it's-there-now-it's-not blur of enterprise.

One result of this constant motion is that solids aren't always as solid as we think they are, and molecules that you thought were pretty consistent sometimes unexpectedly change from one phase

The heat required to raise the temperature of 1 gram of water by 1°C is called a calorie. Don't confuse that with the calories listed on a nutrition label; when you "burn calories" in your daily workout, you're actually talking about kilocalories (or kcal, each one equaling 1,000 calories). One joule equals about one quarter of a single calorie.

to another. For example, some of the molecules in a block of ice will, even if kept below freezing, change into a liquid phase, then usually freeze again. Even stranger, if left alone, the ice will eventually evaporate, as frozen molecules literally boil into vapor in a phase-skipping process called sublimation. Similarly, in a gas, a few cooler molecules will spontaneously combine to form a liquid, then return to gas again, in a continuous state of transmutation.

The most you can say about any material is that it tends toward a particular phase (solid, liquid, or gas) at a particular temperature and a particular pressure—for pressure, too, has a huge effect on phase. Increasing pressure increases temperature—that's why a bicycle tire gets warmer as you pump it full of air—but it also changes the melting or boiling point of a substance. Water boils at a lower temperature on top of a mountain than in a valley, as there is less air pressure to hold it in liquid form. Take a liquid far higher, into the low-pressure vacuum of space, and it instantaneously vaporizes, expands, cools, and then sublimates into tiny shards of crystal.

One thing is for certain, though: However matter changes, heat is never actually gained or lost; it is simply moved from place to place, or transformed from one kind of energy to another. This is at the heart of the laws of thermodynamics (a fancy way of saying "how heat moves"). Just as a gas always expands to fill its container, or people spread out to fill the space in an elevator, heat always spreads into cooler areas. Thus, when you use a thermometer to take your temperature, your body heat cools down a tiny bit as energy moves into the probe, until the two (body and device) are the same.

> **Each degree on the** Fahrenheit scale equals 5⁄9 of a degree on the Celsius scale. That has the curious result that both scales converge at −40 degrees: −40°C = −40°F.

Making Things Cold Few people understand how a refrigerator—or an air conditioner, for that matter—makes the air cool. You can't just take air and "add cold" to it; you need to suck the heat out of the air you already have. When you use a spray can, you notice something curious: The longer you spray, the colder the can gets. The reason is simple: The pressure inside the can drops when you press the button, and the lower the pressure, the less the molecules bounce around, and so the colder the gas becomes. Of course, after a moment, the can absorbs heat from the air around it, so the effect fades.

A refrigerator captures this same effect but carefully controls it—and keeps the chemicals in a closed loop in order to use them again. A substance such as liquid carbon dioxide or Freon is released from a compressed tube, through a spray nozzle, into a larger set of tubes, where the gas rapidly expands and becomes very cold, very quickly. Air from inside the refrigerator is blown over these tubes, transferring any heat to the colder-than-ice gas. The air circulating around your food (or room) gets colder, and the now warmer gas is then pumped out, compressed until it becomes very hot, and run through condenser coils, where all the heat is released into the room (or outside). By the time the gas gets to the end of these coils, it has returned to room temperature (and much of it has turned to liquid),

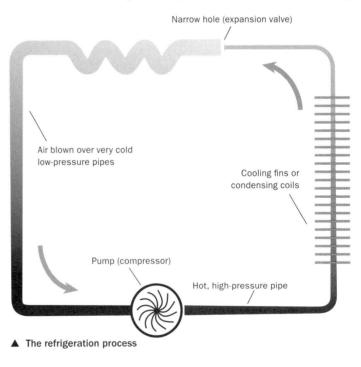

Narrow hole (expansion valve)

Air blown over very cold
low-pressure pipes

Cooling fins or
condensing coils

Pump (compressor)

Hot, high-pressure pipe

▲ The refrigeration process

but it's still highly compressed, ready to repeat the process all over again.

Meanwhile, while you're cooling your food or home or whatever, you're also pulling water out of the air—dehumidifying it. The cooler the air, the less moisture it's able to contain, so water vapor collects (called condensation) around the cold pipes—just another phase change due to heat.

The funny thing about water (which makes up the vast majority of us, the food we eat, and the surface of our planet) is that it expands slightly when cooled. This is unusual, as most other substances get more dense as they cool and freeze. This property of water is one reason life does so well on Earth—after all, the bigger, less dense ice floats instead of sinking, letting water freeze from the top down, so fish and other plants can live protected from the cold.

Unfortunately, even though ice takes up only 9 percent more space than water, that's enough to cause massive destruction where you least want it. When water leaks into a tiny crack and freezes, the force of the expansion can split rock or concrete; metal pipes can burst, glass can shatter. Organic materials like meat and vegetables—which you'd expect to be more pliable—often fare the worst, as sharp ice crystals rip open delicate cell membranes, rupturing their contents. The result, when thawed, is the flavorless, mushy mess that disheartens so many hopeful cooks.

In 1923, the inventor Clarence Birdseye found that quickly freezing a thin layer of fish filets created much smaller and more evenly spaced ice crystals that avoided most of the cell damage. What works for fish works for other food, too, and by 1928, Americans were buying over a million pounds of frozen foods each year. In theory, flash-frozen food will hold indefinitely, but as we've learned, even ice cubes evaporate, and moisture slowly wicks away wherever cold air can get to it, causing unsightly (and tissue-damaging) freezer burn.

Of course, if you want to freeze something really fast, you put it in an environment far colder than normal ice. Take frozen carbon

Because cold air contains less moisture, mountain ranges in temperate areas receive far more snow than the north or south poles. The Antarctic is essentially a desert, with extremely dry air and even less precipitation than Phoenix, Arizona!

Denizens of cold climes know that stepping on snow causes little noise at air temperatures near 0°C, when a thin film of water lubricates the rubbing between the ice crystals. At much lower temperatures there is no water film, so the friction produces a relaxation oscillation called a squeak.

dioxide (CO_2), for example, otherwise known as dry ice. The French scientist Charles Thilorier was the first to change the phase of this gas, in 1834, by placing it under incredible pressure (thus radically raising its boiling point), then releasing the pressure, dropping the temperature so quickly that it desublimated directly into a solid. It sounds like a simple process, but at the time, these kinds of experiments were dangerous; one of Thilorier's assistants lost both legs when the apparatus exploded during testing.

You can buy dry ice at a grocery store, but don't touch it: The surface is −79°C (−110°F), so cold that it burns, destroying your skin

Element	Melting Point (°C) (solid to liquid)	Boiling Point (°C) (liquid to gas)
Helium	−272	−269
Hydrogen	−259	−253
Oxygen	−223	−183
Nitrogen	−210	−196
Chlorine	−101	−35
Carbon dioxide	−78	−57
Mercury	−39	−357
Bromine	−7	59
Phosphorus	44	280
Lead	328	1,740
Aluminum	660	2,467
Silver	961	2,212
Gold	1,065	2,807
Iron	1,535	2,750
Tungsten	3,422	5,555
Carbon	3,550	4,827

▲ Phase change

cells. However, it'll keep a container cold for quite a while as the solid CO_2 slowly, dryly evaporates back into a somewhat harmless gas.

The Hunt for Zero Frozen carbon dioxide's melting point of –79°C is pretty nippy, but in 1983, at the Vostok Station in Antarctica, the thermometer outside dropped to –89.2°C (–128.6°F)—currently the coldest naturally occurring temperature ever recorded on Earth. There have certainly been events even colder, but without the benefit of humans to record them.

At –100°C, rubber tires freeze—a point well taken by recyclers, who shatter them into tiny shards to be reused in other materials. About 80 degrees colder, the oxygen and nitrogen we breathe liquify; and only 40 degrees colder than that, at –219°C (–362°F), they turn solid.

Liquid nitrogen and oxygen have a wide array of uses, as they freeze almost anything virtually instantaneously, perfectly preserving it in stasis. Cryobiologists commonly keep sperm cells, stem cells, and many other plant and animal tissues at –196°C (–320°F) indefinitely with little loss when thawed. And if you can freeze a cell, why not an organ, or even a whole body? In the early 1960s, a Japanese researcher froze a number of cat brains for days, weeks, and even months. After they were carefully warmed with a bloodlike substance, there were measurable (though brief) brain signals that were very similar to those from the original live brain.

Following this promising (though disquieting) evidence, cryonics experts have frozen hundreds of human bodies in liquid nitrogen in hopes that someday technology will advance enough to reanimate these people. Some customers choose to preserve only their brains (removed as quickly as possible after the moment of death), assuming that any future civilization capable of thawing a whole body could just as likely transplant the brain into a new body, or perhaps even create a new brain using the original, intricately woven neurons as a model.

Another benefit of compressed liquid air is that, when released, it boils from a compact form to a gas extremely rapidly—even explosively. In 1926, the American physicist Robert Goddard found a way to control this reaction, turning it into a fuel to propel a rocket—the same technique NASA and every international space agency has used to launch satellites, place astronauts on the moon, and boost shuttles to the International Space Station. Clearly there is great power in carefully managing the cold.

Space can get far colder than our temperate little planet, of course. The surface of Pluto is about −223°C (−369°F), and a crater on the dark side of the moon has been measured at a few degrees colder than that. While it would seem like there would be no temperature at all in the dead of space, far out between the galaxies where no stars burn, astronomers have discovered that even "emptiness" has an amazingly consistent "background radiation" temperature of about −270°C, or −455°F—more easily notated as 2.7 kelvins. This ubiquitous but mysterious heat has fascinating implications for our understanding of how our universe was born and grew.

Shortly after the big bang (approximately 13.7 billion years ago), the universe contained a lot of material in a still relatively small space and glowed unbelievably brightly with extreme heat. Astrophysicists believe that the faint background radiation, which they can detect no matter where they point radio telescopes, is the remnant of those early days, like the heat on an oven's walls long after the oven has been turned off. Curiously, on closer inspection, they've found that space is a tiny bit warmer in some places and a tiny bit cooler in others—for example, it may be 2.7249 K in one spot and 2.7250 K in another. These infinitesimal differences are likely the result of quantum fluctuations that have been stretched out by the inflation of the space-time continuum. In other words, as the universe has cooled and expanded, the massive differences between scorching and less-scorched spots that must have existed within the explosive fireball have all cooled and almost evened out, creating this pattern.

These temperatures are cold enough to freeze almost anything except helium, with a melting point of 0.95 K (–272.2°C, –458°F) — which scientists long considered the final frontier. Then, in 1908, the Dutch physicist Heike Kamerlingh Onnes accomplished the task in an experiment that took seven years of preparation, followed by thirteen hours of slowly cooling the helium until it solidified.

▲ The coldest known natural temperature in the universe is the Boomerang Nebula, a bow-tie-shaped expulsion of gas from a dying star about 5,000 light-years away in the constellation Centaurus. You'd expect an explosion like this to generate huge quantities of heat, but in fact the extremely rapid expansion of the gas into space had just the opposite effect, acting like the expansion tube in a refrigerator and leading to a temperature of about 1 K.

Referring to the explorers who were attempting to reach Earth's poles during that same decade, Kamerlingh Onnes explained his passion for this work: "The arctic regions in physics incite the experimenter as the extreme north and south incite the discoverer."

Where It All Gets Wacky Reaching toward absolute zero is like one of those horrible nightmares where the faster you try to run, the slower you get. We know that zero kelvin (−273.16°C) is a hard stop we can never achieve—a physical extremity like the speed of light. At absolute zero, all atomic and subatomic motion would stop and all particles would come to rest with zero energy. As far as our current understanding of science can see, accomplishing absolute zero would violate one of the fundamental laws of quantum physics, Heisenberg's uncertainty principle, which states that we cannot pinpoint the exact location and momentum of every particle, even electrons.

This kind of boundary—where each step closer is exponentially more difficult—is called an asymptotic limit. In this case, as you get closer to zero, that is, as you are trying to remove more and more heat, you actually start generating heat. But the challenge is worth the effort, for the world of the ultracold reveals some astonishing phenomena.

To play in this realm, you need more clever ways of reducing heat besides just compression and expansion. For example, common evaporation lowers surface temperature, which we humans take advantage of by sweating on a hot day. You can also suck heat out of a system by running a small electrical charge to create a difference in temperature between two metal plates, a device used in many portable camping coolers. But when working at extremely low temperatures, below 1 K, nothing compares to cooling atoms with lasers.

Laser cooling techniques, developed in the mid 1980s with names like Doppler or Sisyphus cooling, all work by focusing two or more intense beams of light at a tiny group of atoms. By precisely tuning

Temperatures

Absolute hot (Planck temperature) 1.42×10^{32} K

Melting point of hadrons into quark-gluon plasma 2 trillion K

Everything 1 second after big bang 10 billion K

Thermonuclear weapon peak 350 million °C

Sun's core 15 million °C (27 million °F)

Lightning bolt 28,000°C (50,000°F)

Center of Earth 6,650°C (12,000°F)

Surface of sun 5,500°C (10,000°F)

Filament inside light bulb 2,500°C (4,600°F)

Natural gas (methane) flame on a stovetop 1,200°C (2,200°F)

Lava 1,100°C (2,000°F)

Wood fire 900°C (1,650°F)

Draper point (where almost all solid materials begin to visibly glow) 525°C (977°F)

Melting point of lead 328°C (621°F)

Kitchen oven 288°C (550°F)

Book cellulose-based paper burns 233°C (451°F)

Water boils 100°C (212°F)

Hottest shade temperature recorded on Earth 58°C (136°F)

Human body temperature 37°C (98.6°F)

Room temperature 20°C (68°F)

Water freezes 0°C (32°F or 273 K)

Mercury in a thermometer freezes −39°C (−38°F)

Coldest temperature recorded on Earth −89°C (−129°F)

Alcohol freezes −114°C (−173°F)

Gasoline freezes −150°C (−238°F)

Boiling temperature of oxygen −183°C (−298°F)

Temperature on Neptune −220°C (−364°F)

Coldest spot on moon −228°C (−378°F or 45 K)

Cosmic microwave background 2.725 K

Coldest natural temperature known (Boomerang Nebula) 1 K

Coldest measured temperature 100 pK

Absolute zero 0 K (−273.16°C or −459.67°F)

Note: Many values here are approximate, as temperatures can vary.

the electromagnetic wavelengths of the lasers, scientists can draw atoms in one direction or another by bombarding them with photons. One of the inventors of this method, the Nobel Prize–winning American physicist Carl Wieman, described the process as "like running in a hail storm so that no matter what direction you run the hail is always hitting you in the face . . . So you stop."

In the 1980s, scientists achieved thousandths of a kelvin. In the '90s, they slowed down the atoms even more, lowering the temperature to millionths of a kelvin, then hundreds of billionths, offering an unprecedented glimpse at the world of the supersmall. Luis Orozco, a physics professor at the University of Maryland, explained in a *NOVA* documentary: "It is as if I were to ask you, 'Could you tell me something about the handles of a car that is passing on a highway at 50 or 60 miles an hour?' Definitely you won't be able to say anything. But if the car is moving rather slowly, then you would be able to tell me, 'Oh yes, the handle is this kind, that color . . .' At room temperature an atom is moving at roughly five hundred meters per second [about 1,100 mph]. However, if I slow it to a temperature that we can now achieve without much work in the lab, two hundred microKelvin, then the atoms start to move about twenty centimeters per second. Compared to something that's rushing in front of you, you'd be able to look at a lot of the details, a lot of the internal structure of that atom."

But it turns out that supercooling atoms encourages them to behave bizarrely—behavior that not only provides insights into the nature of matter but also may allow us to improve technology in extraordinary ways. For example, while some metals are better than others at conducting electricity, some materials, such as lead or buckminsterfullerines, become superconductors when cooled to extremes. A superconductor doesn't just allow electrical impulses to pass through it; it does so without offering any resistance—a current running through a loop of superconducting wire will never fade.

It's unclear how superconductors can pull off this feat of perpetual motion, but it appears that as the temperature drops, the atoms

> **If the sun went out, Earth's** surface would cool to about −220°C, warmed slightly from heat coming from its core.

vibrate less and electrons can slip through more easily. The electrons actually seem to group into pairs, each tugging the other forward, when normally they would repel.

However it's accomplished, superconductivity has led to incredibly powerful and precise magnets—magnets that today power MRI scanners and particle accelerators. Magnetically levitating (maglev) trains based on superconductor technology are still being tested, but they have already broken the world record for fastest-moving train, on an experimental track in Japan, at 581 km/h (361 mph). Someday, whole power grids may be based on superconductors, as estimates suggest that, in transferring electricity, 110 kilograms (250 lb) of superconducting wire could replace 8,100 kg (18,000 lb) of copper wire.

A second characteristic of supercooled atoms is that they can become superfluid—that is, at a certain point, the atoms in liquid helium begin to ignore friction. If you swirl a superfluid, it keeps swirling forever; if you spin its container, the superfluid inside remains motionless. A superfluid can escape through extremely small pores that would normally hold any liquid. Weirdest of all, because a superfluid does have surface tension—like all liquids, it has a slight attraction to the sides of a glass container—it gradually creeps up the sides of a cup until it flows out on its own accord, like a translucent creature that somehow knows how to escape confinement.

But these superpowers were just a small taste of the wonders scientists were about to find in the nanokelvin zone. In 1995, when the physicists Carl Wieman and Eric Cornell cooled a small collection of rubidium atoms to these extremes, they encountered a breakthrough moment: The atoms suddenly shifted phase—not to a liquid, or a solid, but to an entirely new state of matter, never before seen.

To understand this new state, we need to look back to 1924, when Albert Einstein and the Indian physicist Satyendra Nath Bose theorized that as individual atoms neared absolute zero, they would change in an extraordinary way. Quantum mechanics states that all atoms can be described as either particles (things) or waves (energy). Near zero, the theory went, atoms should begin to act less like particles and more like waves; then the waves would get longer until they overlapped, suddenly acting like a single wave—that is, as though all the atoms were one "superatom."

This new state, a "holy grail of cold" called the Bose–Einstein condensate (BEC), is what Wieman and Cornell had created in their lab. Like everything else about quantum physics, BEC is a counterintuitive mystery. The atoms still exist, yet they have expanded their size—their awareness, as it were—in a way that we still don't understand.

The condensate behaves unlike any other material. The atoms vibrate—barely—in unison, a quantum lockstep that acts like a giant magnifying glass on what is usually far too small to see. In 1998, the Harvard physicist Lene Vestergaard Hau found that she could shine a laser into a BEC made of millions of sodium atoms and slow down the light to 68 km/h (38 mph)—a huge leap from its normal speed of 300,000 km (186,282 mi) per *second*. A few years later she found a way to tune the BEC using lasers of specific wavelengths, letting her literally stop the light entirely, then release it again on its way.

Here's how it works. The light pulse is converted to a hologram inside the condensate, literally creating a copy in matter, which can actually be transferred from one BEC to another nearby, like handing over a packet of information, before transforming back into light again. The implications are staggering and point to future quantum computers that may run on light instead of electricity.

On the other hand, these tiny condensates can also explode in an unexpected, extremely tiny version of a supernova—which scientists, recalling the 1960s Brazilian music boom, call a "Bosenova."

What happens in environments even colder than the nanokelvin? We're still finding out. After the Nobel laureate Wolfgang Ketterle trapped a cloud of sodium atoms in place with magnets in 2003, his team at MIT was able to laser-cool the gas to 500 picokelvins—half a billionth of a degree above absolute zero.

The physicist Juha Tuoriniemi at the Helsinki University of Technology's Low Temperature Laboratory has taken a small piece of rhodium metal as low as 100 pK (1×10^{-10} K), but every supercold researcher today is focused on the next breakthrough: the femtokelvin, millions of times colder than the temperatures required to build a Bose–Einstein condensate. Scientists are hungry to see what surprises await in this often unpredictable realm where our everyday assumptions are superseded by the improbable results of quantum physics.

Some Like It Hot It's a common misconception that heat rises. In reality, dense things sink and less dense things get pushed out of the way—which typically means they float up, like bubbles in a drink. That holds true whether it's air in a room or lava under Earth's crust. The difference in density is caused, of course, by heat. Add heat and most substances expand, lowering their density as the atoms and molecules dance and twitch. (We've already seen that ice offers one exception to this; silicon is another. But they are truly oddities among the vast majority of materials.)

Solids aren't in any position to move much, but everyone knows that running hot water over the metal lid of a tightly sealed jar makes it easier to open—the heat literally expands the metal, even if just a small amount. Engineers using steel in railroad tracks and bridges have to take this effect into account anywhere the ambient temperature is likely to rise or fall significantly. On a hot day, a beam may expand several millimeters in length, buckling if an expansion joint hasn't been provided.

A liquid or a gas offers plenty of latitude to move about, creating convection currents in which colder areas drop, get warmed, rise, lose some of their heat, and then drop again. This cycle is particularly helpful in cooking, but we can see it everywhere on Earth: weather patterns, the circulation of the ocean, hot soot rising up a chimney flue. Clearly, heat causes a lot of motion at the macro as well as the micro level.

At some point, if you add enough heat, you can force even more changes. A liquid boils, forcibly rending molecules apart until they fly into a gas. Even solids can change dramatically. If you place an intense heat source under a piece of paper, those pressed-flat plant fibers will undergo a radical transition: Around 150°C (300°F), the cellulose material starts to decompose, releasing into gases. We typically call this mixture of hydrogen, oxygen, and tiny carbon particles smoke. The more particulates that are released, the more smoky it appears. Some substances in paper don't burn without far more heat, of course, so some material remains, darkened, called "char."

If you apply even more heat, an astonishing thing happens: The various molecules in the paper and gas get so excited that they break apart into atoms, which quickly recombine to form carbon dioxide, water vapor, and other molecules. These blindingly fast chemical reactions have an interesting side effect: They generate even more heat, so even if you remove the original heat source, the new gases are so hot that they cause even more molecules to break up. As long as you have fuel to burn, and oxygen for it to react with, the heat keeps the process cycling. Obviously, we know this amazing chain reaction by a simple name: fire.

Fire has been held as magical for millennia, a gift to humankind so extraordinary that it must have been stolen from the gods, and assumed to be so fundamental that the ancients gave it elemental status alongside earth, air, and water. Now we know that fire is simply the transition from one state to another, and the flames we usually see are simply the gases and particles glowing with incandescent heat.

In fact, anything will glow when it gets hotter than 525°C (977°F), named the Draper point, after the nineteenth-century American chemist John William Draper, who first wrote about this effect. (Coincidentally, Draper was also fascinated by photography and is credited with producing the first clear photographs of a female face and of the moon.) To be accurate, anything cooler than 525°C glows,

The science-fiction author Ray Bradbury called his 1953 classic, about a man who burned books for a living, *Fahrenheit 451*. Of course, he could have named it the metric equivalent: *Celsius 233*. This is the temperature at which paper made from wood pulp begins to burn.

too, but with infrared light, so we can't see it. But at the Draper point the light waves carry enough energy that we begin to see a faint red glow. Heat something up to 725°C (1,337°F) and the color becomes positively luminous—red hot, as we call it.

It's easy to remember that red hot is about 1,000 K. At 3,000 K, a material glows bright orange; at 6,000 K, it turns yellow-white. Guess the surface temperature of our sun . . . right, just about 5,800 kelvins. If the sun's surface were hotter, it would appear bright white, or even—at 10,000 K—blue. This color spectrum is called black-body radiation, and it describes how thermal energy gets partially converted into electromagnetic energy, photons that will travel through space indefinitely. As long as there is movement, there is heat, and where there is heat, there is light.

Granted, some flames burn hot but we can't see them—or can barely see them. A pure hydrogen fire burns clear as the gas combines with oxygen in the air and turns into water vapor; a pure ethanol fire burns so hot that the blue is almost indistinguishable in the bright light of day. Plus, different chemicals release different colors when heated, explaining the rich tonal range of a wood fire. And sometimes the color of a flame doesn't tell the whole story: The blue section near the base of a match flame is technically hotter than the yellow tip, but due to a number of real-world factors (like air flow), it's generally easier to light a candle using the cooler tip.

Absolute Hot A candle flame is plenty warm enough for most of us, but it's only the beginning when it comes to the spectrum of hot. Because heat is just another form of energy, you can raise an object's temperature in all kinds of ways, from running an electrical current through it to blasting it with microwaves.

One way to get a gas hot is by compressing it, which also drops its boiling point so that it may return to a fluid state. If you apply enough pressure and heat, you create something called a supercritical fluid—not different enough to earn status as a new form of matter but nevertheless possessing some very cool properties.

"By convention sweet is sweet, by convention bitter is bitter, by convention hot is hot, by convention cold is cold, by convention color is color. But in reality there are atoms and the void. That is, the objects of sense are supposed to be real and it is customary to regard them as such, but in truth they are not. Only the atoms and the void are real."

—Democritus, Greek philosopher

For example, a supercritical fluid can dissolve materials like a liquid and also pass through semiporous solids like a gas. If you infuse a bunch of green coffee beans in a high-pressure bath of nontoxic hot carbon dioxide (CO_2), the supercritical fluid seeps through the beans, absorbing caffeine and drawing it out. Then release the pressure, and the CO_2 suddenly vaporizes into a steam, leaving the decaffeinated beans ready for roasting. Supercritical fluids are used in dry cleaning, essential oil extraction, dyeing materials, and all sorts of other applications.

However, if you warm up a gas even further, you do, in fact, achieve a new, fifth form of matter: plasma. At first, you may not notice any difference between a gas and a plasma, but the latter contains a bunch of atoms that have ionized—they've gotten so excited that they've broken free from their molecular relationships and stripped off some of their electrons, like partiers throwing caution to the wind.

This high-temperature concoction of positively charged ions and negatively charged electrons displays some interesting characteristics. To start with, you can run an electrical current through it, which is what makes neon signs, plasma televisions, and fluorescent lightbulbs glow. Switch on the power, and the molecules in the gas are turned into a superhot plasma. Fortunately, the gas is under such low pressure (there aren't that many atoms buzzing about) that the total heat enclosed in the lamp isn't hot enough to melt things around it.

In these applications, the color we see doesn't come from the heat that the lamps generate; instead, the plasma causes phosphorescent chemicals painted on the glass to luminesce. But in some other instances, the gaslike substance itself lights up as atoms and electrons reunite, resulting in bright light and intense heat. Plasma cutters, which spray a high-speed stream of electrified gas plasma through a nozzle, can cut through steel up to 6 inches (150 mm) thick.

We all know of another common plasma: the sun. In fact, all stars are made of plasma. And, weirder, most of the free-floating gas sparsely spread between planets and stars is in a plasma state. Astronomers estimate that as much as 99.9 percent of all the visible matter in the universe is plasma.

But just because something is superhot doesn't mean it's going to be plasma. Earth's core is hot—at 6,650°C (12,000°F), it's hotter down there than the surface of the sun—but gravity tugs on each atom across thousands of kilometers, creating intense pressure in the center, which is very hot but solid. The larger the mass, the more heat is generated, so Jupiter's core temperature is estimated to be as high as 20,000°C (36,000°F).

Yet that's a cold shower compared with what happens inside the blast of an atom bomb or nuclear reactor, where temperatures of millions of degrees can be achieved through fission—the splitting of heavy atoms like uranium into smaller ones. And fission, in turn, is somewhat pitiful compared with the universe's real power: fusion.

When a huge quantity of hydrogen and helium gas floating in space finds itself collected by gravity into a small enough ball, the atoms begin to smash into one another more and more rapidly. When the pressure and temperature are great enough, say, about 10 million K (about 18 million °F), the hydrogen atoms fuse together, nucleuses joining, creating a new helium atom. Sounds simple, but in the process a lot of energy is released—where "a lot" can be defined as mind-boggling, flabbergasting quantities that result in a fireball called a star.

The sun's core registers at about 15 million K (27 million °F), where hydrogen is consumed at about 600 million tons *per second.* That means the sun, which is only about 4.6 billion years old, is currently in its midlife, about halfway through burning out.

Fusion is not impossible on Earth, however. That's how the H-bomb works: A fission-based atom bomb blows inward on a small bit of prepared hydrogen, causing such intense heat that it fuses. The

Revelation 21:8 implies that Hell has lakes of brimstone (sulfur). At sea level, sulfur boils to gas at 444.6°C (832°F). However, far underground, at extremely high pressure, sulfur can stay liquid as high as 1,040°C (1,904°F).

▲ Large Hadron Collider

results, at temperatures reaching over 100 million K, are devastating. That said, if we can control fusion—or even better, create fusion at temperatures that don't require radioactive explosions to ignite—we would have a never-ending supply of energy. It's a deeply tempting prospect, one on the forefront of many a scientist's mind these days.

Just as Jupiter outsizes Earth, many other stars dwarf our sun. A spectrographic analysis of the heavens reveals stars that have surface temperatures 50,000 times greater than the sun, and cores as high as 2 billion K. In theory, stars could get even larger and hotter, but these nuclear furnaces create something more than just heat and light, something very tiny but that can limit the temperature a star may reach: neutrinos.

Neutrinos are so unbelievably small and slippery that they can travel at near the speed of light through virtually anything. Each second, the sun releases about 2×10^{38} of these little guys (that's 200 trillion trillion trillion neutrinos), and about 65 billion of them end up passing through each and every square centimeter of Earth. That's trillions of neutrinos passing through you right now. It doesn't matter if it's nighttime and the sun is shining on the other side of the planet; neutrinos can literally cruise through Earth, meandering among the atoms, unaffected and unaffecting.

Neutrinos are small, but they do carry a little bit of energy away from a star, and as a plasma reaches about 4 billion K, the atoms become so energetic that the neutrino production actually begins to cool the star significantly. However, if a star becomes massive enough, and hot enough to reach about 6 billion K, the heat triggers such a massive release of neutrinos that the star collapses and then explodes into a colossal supernova. So 6 billion K effectively sets the maximum temperature for a star.

During a supernova, though, things can get crazy and all bets are off on the temperature scale. In 1987, astronomers witnessed a supernova in the Large Magellanic Cloud (one of only two galaxies close enough for us to see with our unaided eye). By careful analysis, they determined that the temperature inside the explosion reached about 200 billion K.

So was that the hottest thing in the universe? Far from it. In fact, you can find a hotter spot just an airplane ride away.

Like obsessive psychoanalysts, many physics researchers insist that the only way to truly understand our universe is to look back into its infancy, to see what crazy things happened within the first second after birth. Back then the heat must have been unbelievably greater than a puny supernova, somewhere in the trillions of degrees. At that temperature, atoms should not only shed their electrons, not only break into their constituent protons and neutrons, but also literally melt into a plasma of quarks and gluons—a seething, primordial broth of elementary particles.

With that in mind, physicists at the Relativistic Heavy Ion Collider at the Brookhaven National Laboratory on Long Island, New York, have been accelerating heavy gold ions around a huge underground ring, speeding the ions up to 99.99 percent of the speed of light, and smashing them into one another. The result is an enormous quantity of heat in a very small space, and in 2010 they achieved a record-setting 4 trillion K (over 7 trillion °Fahrenheit). The experiment confirmed the science, creating a quark-gluon plasma. However, the scientists were surprised to find that the result was more like a "quark soup" than a gas; subsequent calculations indicated that a million times more heat would likely be required to boil it.

On the outskirts of Geneva, Switzerland, CERN's Large Hadron Collider is currently attempting to do just that. Scientists have already achieved trillions of degrees by smashing heavy lead atoms together; how long until the quadrillion- or quintillion-degree mark is smashed, too?

There is, nevertheless, a theoretical upper limit on the thermometer. We know that the hotter particles get, the faster they move. But Einstein also figured out that as particles approach the speed of light, they also increase in mass. Keep increasing the temperature, and at some point each particle of matter would become so dense that it would collapse into its own black hole, causing a minor disruption in . . . well, pretty much everything. The German physicist Max Planck calculated that this would happen at about 1.4×10^{32} K. That's 140 million million million million million degrees. And that, at least in this universe, is absolute hot.

▲ Max Planck

The Creator and Destroyer It is no surprise and no coincidence that virtually every religious tradition describes a divine life-giving warmth that cleanses and sanctifies—but that can also punish or annihilate. For heat is the creator, and every atom in your body and beyond was fused in the fiery depths of a star, often at the moment of its own supernova. And just as surely, heat is the destroyer, rending the elements apart, ending one form and transmuting it into another.

Heat is also the mover and shaker, allowing energy to radiate, infuse, and enable reactions throughout the universe. Without it, molecules could not have bonded together to form the amino acids and other fundamental structures that led to the spark of life, nor could the myriad chemical reactions required to sustain that life—your life—endure.

And yet, watching the ignition of Trinity, the first test of an atomic bomb in 1945—the heat from which melted the New Mexico desert into a crater of radioactive glass 300 meters (1,000 ft) wide—the physicist Robert Oppenheimer recalled lines from the holy Hindu scripture the Bhagavad Gita:

> If the radiance of a thousand suns
> Were to burst at once into the sky,
> That would be like the splendor of the Mighty One...
> I am Death the destroyer of worlds.
>
> —*Bhagavad Gita, chapter 11:12, 32*

Agni, the Hindu god of fire from the 3,500-year-old Rig Veda scriptures, represents the essential life force in the universe, the creator of the sun and stars and the receiver of burned sacrifices, consuming and purifying so that other things may live. Agni may have also given birth to something else: the Latin word *ignis* ("fire"), which begat our English words *ignite* and *igneous* ("having formed from lava").

Nevertheless, heat offers hope, known by any hiker in the wilderness waking to a new sunrise. But removing heat, cooling elements, also offers hope: relief from a scorching daylight, or—in the laboratory—the glimmer of a possibility that we will better understand the building blocks from which we are all made. Cold creates order, though the *very* cold appears to create new disorders that we're just beginning to comprehend.

We are all phoenixes, born from the ashes, living in the radiant glow of the sun—"the force," as Dylan Thomas wrote, "that through the green fuse drives the flower." Heat is the spectrum of life, however you measure it.

TIME

Time keeps on slippin' . . . into the future.

—Steve Miller

THE GREATEST MYSTERY OF OUR UNIVERSE IS A PHENOMENON SO
precious it is considered by some to be holy, but so common that
it is for the most part ignored. More elusive than the slipperiest
substance, yet completely unavoidable, it is the enigma of time. Even
a toddler understands the passing of time, but the most insightful
scientists and philosophers remain baffled as to what it is, how it
works, and how best to measure our place in its ephemeral and
inexorable rhythms.

Of course, a glance at your wristwatch will tell you everything
you probably need to know about time right now, but what does the
ticking of each second really indicate? In the time it takes for you
to read this sentence, our planet will travel 300 kilometers (185 mi)
around the sun, 42 babies will be born, and your laptop computer
could calculate 40 million different chess moves. So what do we
make of the only slightly larger handful of seconds we call our
lifetimes? Truly, in order to understand time, you must let go of your
expectations of what *short* and *long* mean. After all, a millionth of a
second is a long time for a subatomic particle, and a million years is
just a blink of the eye from a cosmic perspective.

The only thing we can definitively say about time is that it
involves change of some sort—without change, there is no time. As
the Greek philosopher Heraclitus wrote, "You cannot step twice into
the same stream." (This is often translated more poetically as "No

man steps into the same river twice, for it is not the same river and he is not the same man.")

When talking about time, it's important to get clear what aspect of time you're discussing. For example, *duration* ("It took one second") is the flip side of *speed* ("In that time, the bullet traveled 300 meters"). But both of those are different from assigning a name to a moment ("At the tone, the time will be . . . noon"). To be sure, discussing time is tricky, but nevertheless an understanding of our universe requires an inspection and a measurement of time.

Measuring Time Sometime in the last ten thousand years or so, we caught on to the regularity of three natural cycles: the sun rises each day, the moon cycles each lunar month, and the sun returns to the same location each year. These simple events, experienced by all of us on Earth, lay the groundwork for the markings on every clockface and calendar.

However, you can imagine the confusion and frustration when early astronomers discovered that the year did not divide by an even number of lunar cycles, or lunar months by days! Instead, each lunar month lasts about 29½ days, and 12 of these lunar months span only 354 of the approximately 365¼ days in a solar year. Trying to make sense of these irregular numbers tried the patience of even the most devout timekeeper.

Nevertheless, the simple 12-month lunar calendar was expedient for the early Sumerian and Babylonian agricultural civilizations, and to this day it forms the basis of the Islamic religious calendar. (That is why the same Islamic holidays show up at different times each year; Islamic dates shift about 11 or 12 days each solar year, returning to the same position about every 33 years.)

A number of other groups, including the Jews and early Greeks, decided to reconcile the solar and lunar calendars, creating a complex and somewhat syncopated system involving 12 years of 12 lunar months interspersed with 7 years of 13 lunar months. On the

π **times 10^7 seconds is a** good approximation of one year (it's 363.6 days). Even better is taking the square root of 10, then multiplying by 10 million seconds (almost exactly 366 days). But π x 10^{16} seconds is about 1 billion years (an eon).

Jewish calendar, every 2 or 3 years there's an extra month, called Adar I, thrown in.

The early Christian church divested itself entirely of the lunar calendar, creating first the Julian calendar and then the slightly more accurate Gregorian calendar, which most of the world uses today. Some say the move was a conspiracy against the feminine aspect of spirituality (as the moon is considered a symbol of women's power); others point out a more practical reason for the church's decision: The farther from the equator you are, the more important the solar seasons are relative to the lunar cycles (which you may not even be able to spot through the cloud cover).

Curiously, the ancient Egyptians had a slightly different compromise to the calendar problem: They rounded the "month" up to 30 days. Twelve of these equal 360 days, a number particularly propitious to mathematicians of the time, who had standardized on the ancient Sumerian sexagesimal (base-60) counting system. After all, if you're primarily concerned with dividing parcels of land, or goods at market, or times of the year, counting by 60 is extremely efficient because it can be divided equally by so many numbers: halves, thirds, quarters, fifths, sixths, tenths, twelfths, and so on. Sixty, and by extension 360, is a beautiful number when you don't have a calculator. In fact, there are still communities of people in Asia who count on their fingers by pointing with their thumb to each finger bone on one hand (up to 12) and then keeping track of sets with fingers of their opposite hand (up to 5), totaling 60.

So perhaps it was natural that if a year were split into 12, then the day and night would also each be split into twelfths, leading to 24 hours from one sunrise to the next. Each hour could then easily be split into 60 minutes and each minute split into 60 seconds, leading to a wonderful symmetry of 360 seconds per hour, like the 360 days per year. It all seemed so perfect, until the Egyptians recognized that the year actually required five or six extra days. It appears they reluctantly snuck these into their calendar, almost as an afterthought.

You know that each four years is a leap year and you have to add a leap day, February 29. And you may know that you have to omit that extra day every 100 years. But do you know that you have to put it back in every 400 years? That's why 2000 was a leap year but 1900 was not. And don't forget to remove the leap day every 4,000 years, or reintroduce it in each century year that, when divided by 900, leaves a remainder of 200 or 600.

Q: In what year was Friday, October 15, the day after Thursday, October 4?

A: In 1582, when Pope Gregory instituted a new calendar!

Earth rotates on its axis in
23 hours, 56 minutes, and 4
seconds (86,164 seconds).
This is called its sidereal
period (which means its
rotation relative to the stars).
But we count a day as the
time it takes for the sun to
return to the same position,
which takes about 4 extra
minutes. That's because Earth
is not only spinning but also
orbiting slowly around the sun.

Internet Time was invented
by the Swatch watch company
in 1998. Also called .beat,
it specifies 1,000 intervals
across a single day. That
means each .beat equals
1 minute 26.4 seconds.
The "zero point" of Internet
Time, called @000, begins at
midnight Central European
Winter Time.

Although the days, months, and years all follow (more or less) astronomical intervals, the seven-day week is perhaps the most cosmic of all, as it's the only measurement that stems entirely from the Hebrew Bible. Exodus 20:9–10 states clearly that "six days shalt thou labor, and do all thy work, but the seventh day is the Sabbath of the Lord thy God." However, other than divine instruction, there's nothing particularly special about a seven-day week, and a number of other cultures standardized on other convenient weekly patterns of four, eight, or even ten days.

A 10-day week has long suited those attracted to the idea of "metric time," where each of the 12 months would be split into 3 weeks of 10 days, each day would contain 10 hours, and each hour would have 100 minutes of 100 seconds each. It sounds preposterous, but of course it's no more arbitrary than the clocks we use today. Metric time gained followers in the French Revolution's flush of reinvention, but the idea—with its deciday, milliday, and microday measurements—soon died out, in part because early nineteenth-century Christians felt its lack of a Sabbath proved the metric system was an abomination.

Since then, rather than reinvent the counting system, scientists have decided to declare ever more precise definitions of time measurements. In 1954, the General Conference on Weights and Measures decided that 1 second should equal exactly 1/86,400 of a mean solar day. Feeling that this definition was still too wishy-washy—after all, the length of a solar day literally shifts during massive geologic events—scientists searched the physical world for a universal standard and alighted on a number that should be reproducible anywhere in the universe: the frequency of microwave light that a heated cesium 133 atom absorbs or emits—specifically, the time it takes for 9,192,631,770 of these electromagnetic wavelengths to pass. True, this definition of a second, which was agreed upon at the 1967 conference, is a value only a scientist could love, but it set the stage for extraordinarily precise measurements of time.

▲ The FOCS1 caesium fountain atomic clock at the Federal Office of Metrology (METAS) in Switzerland

Today, the world's most accurate clocks are white-coat operations, cobbling together lasers, near absolute zero temperatures, and magnetic traps to manipulate the spin on electrons. Physicists on the cutting edge of science can now create timepieces based on quantum logic, accurate to within one second every 3.7 billion years.

This level of precision may seem like overkill to anyone who is just trying to catch a train, but a surprising amount of our science is based on the careful measurement of time. Remember that our basic division of space—the meter—is defined by how far light travels in 1/299,792,458 of a *second*. Similarly, both the fundamental scientific values of the lumen (which measures the amount of visible light from a source, like a lightbulb) and the ampere (or amp, which measures electrical current) are based on the value of a single second of time. Even measuring pressure in the real world relies on measuring force, and force is based on acceleration, which is determined by the passage of time! Without extraordinary clocks, the finest measurements would be impossible.

While scientists focus on measuring ever finer durations, the rest of us concern ourselves with the naming of time—assigning dates to historical events or setting our clocks based on political whims. It's important to remember that every chrononym (a word that refers to some particular time, such as "springtime" or "teatime") is arbitrary; every calendar is based on cultural norms and sometimes curious assumptions. According to Jewish scholars from the Middle Ages, our universe was created in 3761 BCE, which explains why the new millennium was celebrated as 5761 in some quarters. James Ussher, a seventeenth-century Irish Anglican bishop, calculated it slightly differently, noting that creation began the night before Sunday, October 23, 4004 BCE.

Obviously, the Chinese calendar is different from either of these, as are the Islamic and Hindu calendars. And many people put their faith in the Maya Long Count calendar, which prescribes that the fourteenth *b'ak'tun* (numbered 13.0.0.0.0) was to commence on December 21, 2012, signifying either the end of the world or at least the end of some people insisting it'll be the end of the world.

The capricious nature of calendars is equaled only by the somewhat random method of setting our watches. Only 150 years ago, clocks around the world were synchronized at noon by some local official's judgment on when the sun was directly overhead. For instance, in the early nineteenth century, the *Chicago Tribune* regularly reported the variances among fifty-four different local times across Illinois and Michigan alone. This was manageable when you worked almost entirely with people in your own town, but it quickly became untenable as people began to travel. The quick and vast expansion of the railroad, more than anything, created the need for a widespread sense of "official time," and it was the railroad companies that first instituted, in 1883, standard time zones across the United States and Canada.

"O God! methinks it were a happy life . . .

To carve out dials quaintly, point by point,

Thereby to see the minutes how they run—

How many makes the hour full complete,

How many hours brings about the day,

How many days will finish up the year,

How many years a mortal man may live."

—Shakespeare, *Henry VI, Part III*

The following year, at the International Meridian Conference, delegates from twenty-five nations agreed on a worldwide system that outlined 24 fifteen-degree wedges around the globe, each marking an additional hour forward or backward in time. The "zero point" from which all zones are measured is in Greenwich, England— leading to Greenwich mean time (GMT).

In a perfect world, these smooth lines would run from pole to pole like stripes on a beach ball. But a wide spectrum of commercial and geopolitical interests has made these lines ridiculously ragged. The most extreme example is China, which until 1949 contained five different time zones but now maintains one "Beijing time" across the entire nation—even though this means "noon" doesn't come until the middle of the afternoon for cities in western China.

The establishment of GMT also created another arbitrary demarcation: the international date line, which runs more or less through the middle of the Pacific Ocean. When you travel east across

▼ The 24 world time zones rarely follow straight lines. Stripes indicate regions that are ½ hour different.

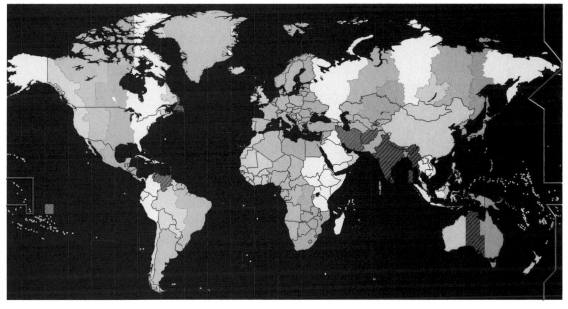

CIA World Factbook

the date line, you jump forward an hour* but back a day. A traveler from Tokyo to Seattle, therefore, experiences the odd sensation of technically arriving before she left.

Here's another example of human-created time line oddities: In 1995, the tiny country of Kiribati, comprising more than thirty small atolls and islands that straddled the international date line, decided to redraw the line. (Some have cynically suggested this was an attempt to increase tourism, as it would suddenly jump to the head of the line as the first country to enter the twenty-first century.) The change means that today, while eastern Kiribati and Hawaii have approximately the same longitude, calendars in Kiribati are bizarrely a day ahead *and* two hours behind Honolulu. Samoa, not content to slouch behind, also recently jumped ship, deciding to entirely skip December 29, 2011, and leap into the future a day.

Imagine an alien arriving on Earth, attempting to make sense of the mess we've made of time. From seconds to light-years, our measurements can be understood only from our provincial, terrestrial perspective. And yet, if we were to try to explain our sense of time to that visitor, we would have to start at the very beginning: the durations and speeds that we as humans can actually sense and comprehend.

Human Time A second, an hour, a year—these are all part of human time, time we recognize, time at the scale in which we live. We have named each aspect of our human time, sometimes with language as flowery as a perfume commercial. It's happy hour, or the witching hour; it's as slow as molasses, or as fast as a bat out of hell. The fastest events happen in just a blink of an eye, or in a flash. That makes sense, as a blink of an eye takes about a third of a second, and our eyes can reliably see events down to about a tenth of a second. Anything faster than that becomes a blur, or doesn't register at

Time zones were widely used across the United States after 1883, but it wasn't until 1918, in the midst of World War I, that Congress passed the Standard Time Act, making these time zones official. This act also instituted the still controversial daylight saving time, which led to its being called "war time." (Ironically, this system had actually been invented in Germany as a fuel-saving measure.) Twice each year, as clocks are set an hour forward or backward, pundits around the world come out to debate this somewhat anachronistic measure. But in 1997, it got ugly: Nearly 2,000 students rioted at Ohio University, because the bars closed an hour earlier than the students expected.

*Actually, in some areas, the date line and the time zone lines diverge, so this isn't always technically accurate.

all. But project a flipbook of static images in succession at twenty or thirty frames each second and they blend together seamlessly, fooling our eyes into believing we see smooth movement. Thus the movie was born.

On the other end of our human sense of time, a very rare event—one with a long duration between occurrences—happens perhaps once in a generation or once in a lifetime. A generation is usually considered twenty to twenty-five years; a lifetime, only a few times that. Scientists have discovered that most animals live, on average, only about 1 billion to 1.5 billion heartbeats. This is why larger animals, with slower and more efficient metabolisms and heart rates, typically live longer than small, fast-heart-beating creatures. There are exceptions, of course. Parrots can live as long as a human, though they have twice our heart rate. And humans have gamed the system with medical advances, so that we now average about 2.5 billion heartbeats.

But who is really living those billion-plus heartbeats? As the roboticist Steve Grand has pointed out, we are not who we were as children, and our earliest memories could almost be ascribed to someone else. After all, our entire blood supply is re-created anew every three months, our skin replaced every two weeks; in fact, every cell in our body is replaced at least once each decade, and there are few molecules in us today that were with us as children! Our sense of self remains consistent throughout our lives, as do our memories, but in truth we are each like a mountain lake, constantly renewing as we empty, the flow maintaining our form (more or less) and warding off stagnation.

Nevertheless, the thing we call our self is living for around a century, give or take. From one perspective, a lot can happen during those approximately three billion seconds; but from another perspective, very little.

A single second—the building block on which all other time is now measured—is the time it takes an average heart to beat once, or the time in which we pass 1 meter (about 3 ft) during a leisurely

"Which is quicker, a jiffy or a flash? I think there are two flashes in a jiffy, myself. But God knows how many jiffies there are in two shakes of a lamb's tail. But why did they use two shakes of a lamb's tail? What's wrong with the basic unit of measurement, one shake of a lamb's tail? We can do our own arithmetic, thank you."

—George Carlin

In the early years of train travel, passengers propelled up to 55 km/h (35 mph) described the experience as "breathtaking."

stroll. But in that one second, a hummingbird flaps its wings 70 times, sound travels 340 meters, and a flash of light rushes past 300 million meters.

A 90 mph fastball takes only half a second to reach home plate, and a batter must decide whether to swing in just a portion of that time. It took Jamaican-born Usain "Lightning" Bolt only 9.58 seconds to run 100 meters in 2009—an unheard-of act, an average of about 10 meters per second (that's 38 km/h or 23 mph).

Of course, these human acts pale in comparison to abilities in the animal world, where the cheetah and sailfish tie the speed records, one running and one swimming, covering 31 meters in a second (113 km/h or 70 mph). The peregrine falcon beats them all, but only with the help of gravity's pull: Nested atop cliffs (or, increasingly, on skyscrapers in large cities), it can dive toward food at more than 90 meters per second (322 km/h or 200 mph).

At a scale too small to detect, oxygen molecules at room temperature zip through the air at over 450 meters per second (1,600 km/h or 1,000 mph). And at a scale too large for us to sense, Earth spins through space at about the same rate—the equator turning at about 1,680 km/h. An aircraft or bullet passing through air that fast would create a sonic boom, but in this case, Earth literally drags the air with it, so we're spared the noise.

Not all planets turn so rapidly; Venus, for example, rotates only 1.8 meters each second (4 mph)—at this rate, you could walk its equator at a brisk pace and appear to stop time, or at least stop the sun's progress across the sky.

That said, most objects in space tend to stretch the limit of our ability to understand speed. In a single second, a communication satellite, traversing the sky 36,000 km (22,000 mi) above sea level in a round-each-day geosynchronous orbit, travels 3,100 meters in a second (11,160 km/h or 6,900 mph). The space shuttle was twice as fast, accelerating from 0 to 17,000 mph (27,360 km/h) in just over 8 minutes. But that's only fast enough to maintain an orbit! To propel a rocket beyond Earth's immense gravitational pull, it must reach

One second is just long enough to pronounce about five unhurried syllables ("one-second-and-one . . . one-second-and-two . . .").

In an effort both to democratize and to settle the tiny differences between solar time and atomic time, scientists created the Coordinated Universal Time (UTC) in 1961. UTC is also called Zulu time and is the worldwide standard in the aviation industry, avoiding confusion by ensuring that every pilot is using the same 24-hour clock.

speeds of over 11,200 meters per second (40,320 km/h or 25,000 mph). That still isn't fast enough to escape the sun's orbit and break away from our solar system—a feat that requires an object to travel almost four times faster.

At the opposite end of our human scale of time, a sloth, infamously slow due to a hypervegetarian diet lacking in protein and fats, can travel only about 10 cm per second (0.2 mph); a snail, limited by its mode of transport as much as its size, slimes along a tenth as fast as that. Although they may strain our patience, we can see these movements, which is not the case for many other phenomena around us. While it varies widely from person to person, hair grows, on average, about 4.5 nanometers per second—that's impossible to see on short time scales but easy to perceive as stubble growing at ⅜ millimeter per day, or about 1.25 centimeters (0.5 inch) per month. You might call these speeds "glacial," but hair grows far slower than glaciers migrate—as quickly as two tenths of a millimeter per second (17 meters or 56 feet per day).

Nevertheless, the speed of a glacier, or hair—or even a flower sprouting, blooming, fading, and dying—are all well within our range of understanding. We can wrap our heads around these human scales in a way that we cannot fathom as we step beyond the century and millennium, or look more closely at events faster than a flash.

Geologic Time Our planet is about 4.5 billion years old, originally formed from floating interstellar rock and ice pulled together by gravity. Imagine that a single year is represented by a 1-millimeter length of string; a century is 10 centimeters (about 4 in.), a millennium is 1 meter. To demonstrate the age of Earth, you would need a string that spans from San Francisco to New York.

Obviously, for those of us who usually consider "long term" as having to do with our retirement, it takes a while to warm up to a perspective from which "soon" means sometime within 20,000 years. After all, 20,000 years compared with the age of our planet is like one minute in five months, or three hours within a human lifetime.

In thoroughbred horse racing, one length is about one fifth of a second.

The oldest living individual organism on Earth is a Great Basin bristlecone pine tree (*Pinus longaeva*) in the mountains between California and Nevada. It is 4,842 years old, based on a ring count of a sample core. This tree, known as Methuselah, was more than two centuries old when the Great Pyramid of Giza was constructed. If you include plants that clone themselves, the longest-living organism may be a single clonal colony of *Populus tremuloides* (quaking aspen) in Utah that has been growing for an estimated 80,000 years.

Here's another way to think about it: You've likely seen time-lapse movies of clouds moving or even seasons changing, but imagine a time-lapse where each frame captures a moment every ten thousand years. If we started shooting at the birth of the planet, the finished movie (to the current day, at least) would be four hours long, and the entire history of the human species wouldn't show up until well into the final one second of footage.

At this rate, what scientists call geologic time, our sense of how the world works begins to break down. For example, over thousands of years, diffusion—the process in which molecules comingle, like gas mixing in a container—affects solids. That's why gold and lead objects found next to each other in Egyptian tombs have merged, as though they have melted together.

On a slightly longer scale, there is no doubt that humans have significantly affected our planet's climate in recent years, but climatologists must also take into account the normal weather cycles, which can be seen only by taking a geologic view. For example, Earth's axis (which points roughly toward the North Star) rotates slightly but returns to the same point every 25,784 years—about as long as it takes a ray of light from the center of our galaxy to reach us on Earth. Think about it: The last time the stars were in the same location they are today, humans were painting in caves during the last ice age.

This normal wobbling is added to a 41,000-year cycle where the tilt of the poles varies between 22.1° and 24.5° and another cycle in which every 100,000 years Earth's orbit shifts from elliptical to circular, then back again. It doesn't sound like much, but a few degrees mean the difference between a lovely spot of weather and an ice age that blankets much of the planet in snow. The irony of our current global-warming problem is that, because the axis tilt is decreasing, we may be heading toward another natural ice age in ten or twenty thousand years, and the warming of the planet may push it back a bit.

"The future is something which everyone reaches at the rate of 60 minutes an hour, whatever he does, whoever he is."

—C. S. Lewis

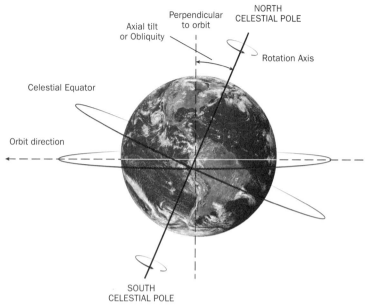

North Celestial Pole

Perpendicular to orbit

Axial tilt or Obliquity

NORTH CELESTIAL POLE

Rotation Axis

Celestial Equator

Orbit direction

SOUTH CELESTIAL POLE

▲ Earth's wobble

Extending our perspective over an even larger span of time, we know that landmasses on floating tectonic plates move at about the speed fingernails grow—South America and Africa are moving apart and the Atlantic Ocean is widening at about 4 centimeters per year. So over the next million years, Los Angeles will creep about 40 kilometers north-northwest of its present location. Even a million years—called a megannum (abbreviated Ma)—isn't that long: If we could build a spaceship that traveled at the speed of light, a million years wouldn't get us to even the halfway point on a journey to the Andromeda galaxy.

Scientists who think in terms of megannums like to mark off the ages of Earth's development, like lines on a doorframe as a child grows. In case you're taking notes, at the time of this writing, we're in the Phanerozoic geologic eon (which started about 542 Ma

Earth's rotation is slowing, due to the gravitational pull of the moon on the tides, leading the length of the day to increase by about 2 seconds every 100,000 years.

ago), in the Cenozoic era (which began when the dinosaurs died off, about 65 Ma ago), and—for the last 11,000 years—the Holocene epoch. Some argue we're also in a new era, called the Anthropocene, characterized by a vast array of human-created sediments spread about the planet's surface (plastic and other refuse, mostly).

It's odd to think about, but sediment and waste—what animals and plants create during their short lives and leave behind afterward—actually make up a huge portion of Earth's outer layers, from rock we walk on to the atmosphere we breathe. For hundreds of millions of years after the creation of Earth, the planet was a superheated ball of sterilized rock and gas, bombarded with asteroids, devoid of any life. But about 3.5 billion years ago, it cooled down enough to support shallow pools of water, in which complex molecules combined to form tiny single-celled bacteria. The bacteria lived, divided, and died, leaving behind their microscopically small shells. A few trillion would have made no difference, but trillions of trillions, over hundreds of millions of years . . . well, it all adds up, and the sediment slowly built up in layered mats that today can be seen as stromatolite rock.

Then, about 2.5 billion years ago, as available organic food sources became scarce, the bacteria learned a new trick: Take water, carbon dioxide, and sunlight and convert them into energy. Unfortunately, this clever technique, called photosynthesis, releases a deadly waste product: oxygen. Well, it was deadly to them at the time, just as carbon dioxide (which we exhale) is toxic to us.

Most of the oxygen combined with iron and other molecules, which over time got trapped in thick layers of sediment (what we today mine as iron ore). Eventually, though, about a hundred million years later, all the available iron oxidized and the oxygen had nowhere to go but up into the atmosphere, resulting in the Great Oxygen Catastrophe, during which the majority of life on Earth died. Of course, one person's toxic waste is another person's chance at life, and it wasn't long (remember, we're talking geologic time here) before tiny single-cell and then multicellular creatures appeared

> **"Time is the fire in which we burn."**
> —Delmore Schwartz, "Calmly We Walk Through This April's Day"

A lustrum is 5 years (from the Latin "to wash," as a form of purification after the census was taken every 5 years in Ancient Rome).

that could take advantage of the oxygen-rich air and the brand-new protective layer of high-altitude ozone (a molecule made of three oxygen atoms that stops most harmful radiation from getting through from space).

There is an ongoing debate between creationists and evolutionists, in which the former insist that complex structures (such as the eyeball) could not have evolved—there are too many different but interrelated parts that must work in perfect synchrony. But what most creationists don't take into account is the incredibly long periods of time involved. Of course an eye could not develop in a year, or even a million years. Evolution of plant and animal life—like sediment—happened over a time span that we, as humans, can scarcely comprehend. It took perhaps 400 million years for multicellular organisms to evolve into the simplest animals, like sea sponges or jellyfish; another hundred million years for fish; then insects and reptiles showed up over the next couple of hundred million years. Throughout this time, plants grew huge, developed, died, and—along with the animals—left immense amounts of rich organic sediment that we now know as fossil fuels, such as coal, oil, and natural gas.

That all happened long before dinosaurs appeared, about 240 million years ago. The dinosaurs ruled Earth for a measly 175 million years, then were suddenly wiped out, probably within hours of a six-mile-wide meteor hitting near the eastern coast of what is now Mexico. Once again, when one door closes, another opens, and the animals that survived (the ones that could burrow, swim, and adapt quickly) had a tabula rasa with which to start anew.

Finally, not until about two and a half million years ago—63 million years after the great dinosaur extinction—did the earliest forms of genus *Homo* appear, and not until about 100,000 years ago did our particular species of *Homo sapiens* show up.

Once again, these geologic time spans are so vast that in order to find our place in them, perhaps we should try to compare them with something we know well: our own human life span. Let's imagine

> "Time is the substance from which I am made. Time is a river which carries me along, but I am the river; it is a tiger that devours me, but I am the tiger; it is a fire that consumes me, but I am the fire."
>
> —Jorge Luis Borges, "A New Refutation of Time"

that "Mother Earth" is celebrating her 50th birthday right now—that is, her whole 4.5 billion years were compressed down into 50. In that case, the first signs of simple life didn't show up until she was age 11. The first animals with eyes (such as trilobites and horseshoe crabs) appeared just before her 45th birthday. Dinosaurs went extinct a few months after she turned 49. *Homo erectus* (Java man) learned to control fire this week, and *Homo sapiens* showed up yesterday. The last major ice age ended an hour ago, and civilization—virtually all our recorded history, art, and science—began a few minutes later. Jesus was born a little over 10 minutes ago. And the age of modern computers began 17 seconds ago. Happy birthday, Mom.

Cosmic Time Once you've trained yourself to think in terms of billions of years—assuming that's even possible—you can see patterns of movement and change at a galactic and universal level that is impossible to consider by simply looking at the sky on any given night. That's not to say that astronomical objects are moving slowly! They're just moving slowly compared with much larger things; it's all relative. Within our solar system, for example, the fastest-moving planet is Mercury, named after the Roman messenger of the gods, which orbits the sun at about 48,000 meters per second (about 107,000 mph); Earth is lazy in comparison, managing only about 30 km/s (67,000 mph).

The sun, and our solar system as a whole, is traveling through space, rotating around the center of the Milky Way galaxy at 800,000 km/h (500,000 mph). And while all that is happening, our galaxy is moving through space at over 2 million km/h (1.2 million mph). That all sounds breakneck until you consider that it takes more than 225 million years for us to make a single orbit—one galactic year.

Our sun is about 4.63 billion years old, or about 20 of these galactic rotations. Back when it was created, five eons ago (an eon is 1 billion years), we were all just a nebula of gas, ice grains, and dust floating through space—mostly carbon, oxygen, silicon, and iron generated in the dramatic explosions of earlier stars. Some

> "Science cannot solve the ultimate mystery of nature. And it is because, in the last analysis, we ourselves are part of the mystery we are trying to solve."
>
> —Max Planck, physicist

shock wave from a nearby supernova may have shoved enough of this matter together that it condensed into a star and planets, similar to how dust bunnies clump under your bed. The universe is like that: Stars grow from dust, then live and die, and sometimes blow up, offering life to new stars and planets. It's hardly newsworthy; billions of stars are born and die each year across the vast reaches of our universe.

It's likely that genesis—what some folks call the big bang—happened about 13.7 billion years ago, plus or minus 110 million years. That's almost 14 million millennia—though only about 60 of our galactic years. Consider Carl Sagan's famous "cosmic calendar," in which the history of the universe is compacted into a single solar year. At this scale, if the universe began on January 1, the Milky Way started forming in March, though our sun and Earth weren't born until September 1. Later that month, single-celled creatures appeared, followed sometime in November by multicellular organisms. Mammals appeared on December 26. Dinosaurs were wiped out on December 29. All human prehistory (from the first known stone tools) and history occurred during the final hour of New Year's Eve, and our entire recorded history of civilization takes up just the last 22 seconds of the last minute of December 31.

The amazing thing is that the universe is still relatively young and impetuous, changing all the time. Only a billion years from now, the sun will likely grow so hot that all life is extinguished from Earth's surface. In a few billion years, there's a small chance that fluctuations in the planets' orbits will cause Earth to crash into either Mars or Mercury, and a very good chance that the Andromeda galaxy will smash into ours, forming one incredibly big elliptical galaxy. Our sun will probably survive this upheaval but go on to burn out seven billion years from now. (There's an old science joke about the woman at the back of the lecture hall who, upon hearing this news, nervously asks for a clarification, then says, "Oh, *billions*! Goodness me, I was worried; I thought you said *millions*.")

If you compressed the age of the universe into a single day, the average human life span would take only 1/2,000 of a second.

While there is nothing older than the universe (at least nothing we can measure), scientists often calculate far longer lengths of time, particularly when considering astronomical lifetimes. The estimated life span of a red dwarf star about a tenth the mass of our sun is 312 exaseconds—that's 10^{18} seconds, or roughly 10 trillion years. With that in mind, given our current understanding, the last star in the universe will likely die about 100 trillion years from now. Nevertheless, the universe itself will go on, primarily made up of black holes, which evaporate incredibly slowly due to something called Hawking radiation. One relatively small black hole the mass of our own sun would take about 6.6×10^{50} yottaseconds to burn away— that's 6.6×10^{74} seconds, 2×10^{67} years, or over a trillion trillion trillion trillion times longer than the current age of our universe.

These numbers are staggering, and yet, just as the largest stars gain their size and energy from the smallest atoms, even the longest events are constructed of the thinnest slices of time, one moment after another.

Very Fast Things In his book *Metamagical Themas*, Douglas Hofstadter describes how, in the 1940s, Nicholas Fattu led a team of ten people working full-time for ten months to solve a tricky math problem. Twenty years later, he used a room-sized mainframe computer to find the solution in just twenty minutes (which included discovering errors in the original work). Today the solution could be found even more accurately in less than a second using a laptop computer. But what's really going on in that single second? What would we find if we slowed time, expanding it until we could peek in on events that appear virtually instantaneous?

You can see individual images flickering by at a rate of ten per second; you can hear individual clicks or strums that fast, too. But speed up the film or soundtrack and a magical thing happens: Our brain blends the experiences together, creating an amalgam, a new experience different from the sum of its parts. Within a tenth of a second, a small hummingbird can beat its wings 7 times, creating

Speed

Fingernails grow	1.2×10^{-9} m/s (2.6×10^{-9} mph)
Moon receding from Earth	1.3×10^{-9} m/s (about 1 nm/s)
Average growth rate of child	1.8×10^{-9} m/s (4×10^{-9} mph)
Growth rate of bamboo	6×10^{-7} m/s (1.3×10^{-6} mph)
Garden snail	0.002 m/s (0.004 mph)
Audiocassette tape speed	0.0476 m/s (0.106 mph)
One knot (nautical mile per hour)	1.852 km/h (1.151 mph)
Average walking speed	1.2 m/s (2.5 mph)
World record swimming	2.3 m/s (5.2 mph)
Comfortable bicycling speed	6 m/s (13 mph)
Fastest human running	12 m/s (27 mph)
Horse in gallop	13 m/s (30 mph)
Object falling 10 m	14 m/s (31 mph)
Sky-diver in midflight	54 m/s (120 mph)
Sneeze	up to 46 m/s (104 mph)
Golf ball driven off tee, arrow fired from longbow	60 m/s (130 mph)
Stock car	90 m/s (200 mph)
Wind in tornado	112 m/s (250 mph)
Fastest train with wheels (non-maglev)	160 m/s (357 mph)
.22 long-range centerfire bullet	400 m/s (1,300 ft/sec, or 895 mph)
Earth's rotation at equator	464 m/s (1,038 mph)
SR-71 Blackbird, fastest jet aircraft	981 m/s (2,194 mph)
North American X-15 rocket-powered aircraft	2,020 m/s (4,519 mph)
Satellite in geosynchronous orbit	3,100 m/s (6,900 mph)
Space shuttle, and International Space Station in orbit	7,743 m/s (17,320 mph)
Apollo 10 manned spacecraft, return trip from moon	11 km/s (24,791 mph)
Earth orbits the sun	30 km/s (66,620 mph)
Helios space probe (orbiting the sun)	70.22 km/s (157,078 mph)
Solar system orbiting the Milky Way	216 km/s (483,000 mph)
Milky Way moving through space (relative to cosmic background radiation)	550 km/s (1.2 million mph)
Rotation of fast neutron star	38 Mm/s
Light in a vacuum	299,792,458 m/s (186,282 mi/sec)

0 20 40 60 80 100 120

MPH

a low hum in our ears. Vibrate a string 44 times in the same tenth of a second, and our ears hear A above middle C. In that same split second, the sound will travel 33 meters (108 ft) through the air, but a light flipped on at the same moment travels quite a bit farther—in a tenth of a second, the light could travel three quarters of the way around Earth. If that light wave were an AM radio signal at the lowest end of the dial, it would have vibrated 55,000 times in that flash of light.

It takes two tenths of a second to react to something you see—slightly less for an auditory stimulus—but it takes just a hundredth of a second for a nerve impulse to travel from your brain to your hand. Actually, signals travel at varying rates through different neurons (nerve cells), from 1.6 to 600 km/s (1 to 268 mph), which sounds fast, but it's millions of times slower than electricity flowing through wire. When you touch something, electrochemical impulses send the message to the brain extremely quickly, but if that thing is sharp or hot, you won't recognize that until later, because pain signals travel 100 times slower.

We've become familiar with hundredths of seconds in large part due to the sports we watch. While human reflexes are not fast enough to judge even tenths of seconds, electronic instruments are clearly far more accurate. In the 2008 Olympics, when Michael Phelps won his seventh gold medal for swimming in the 100-meter butterfly, his time was 50.58 seconds, just one hundredth of a second faster than Milorad Čavić. The time was based on an ultrathin plastic touchpad mounted on the wall and was reinforced by high-speed cameras running at a hundred frames per second. Phelps was recorded slamming his hands down on the pad one frame faster than the Serbian.

The very fastest aspects of our lives can be captured at the scale of milliseconds (thousandths of a second). A housefly flaps its wings once every three milliseconds; a normal point-and-shoot camera can record a single photograph in good light, stopping human motion, in one millisecond. In running and bicycle racing, extremely fast

> **"Time is an illusion. Lunchtime doubly so."**
>
> —Douglas Adams, novelist

cameras, shooting one thousand frames per second, focus on the final moment at the finish line. Coaches and athletes will argue, but the difference between gold and silver at this scale—especially when it comes to swimming, running, bicycling, skiing, or horse racing, where the competitor's environment can change uncontrollably from one moment to the next—is a matter of luck as much as skill.

When most people think about financial trading, they imagine people yelling on a crowded market floor, but today the big trades are done not by humans but by silent computer algorithms making split-second decisions based on news collected from around the world. Case in point: A new transatlantic cable is being laid, at the cost of $300 million, in order to shave six milliseconds off the time it takes for a signal to travel from Europe to New York. Many analysts applaud this move, noting that even a single millisecond's advantage means an extra $100 million each year to a large hedge fund.

We can somehow understand milliseconds: A thousand events in a second is shocking but still within comprehension. However, microseconds (μs)—millionths of a second—teeter on the edge of disbelief. A microsecond is to a single second what 1 second is to 11.5 days—in other words, if you took one step each second, you'd have to walk for 277 hours straight to reach a million. To give you a sense of how long a microsecond is, it takes about 500,000 μs to click a mouse. And yet, because sound reaches one of our ears just 600 μs before the other, we can identify where it originated and turn our head to it, even with our eyes closed. (How we manage this is actually a mystery, as the signal from the ear to the brain may take as long as a millisecond.)

The half-life of a neutron in isolation is about 10.5 minutes. That's a billion times longer than the half-life of any other known atomic particle.

At these rates we also begin to encounter subatomic particles. For example, the entire life span of a muon—generated by powerful atomic collisions that transform energy to mass—is only about 2.2 microseconds, after which it quickly ruptures into an electron and two neutrinos. A lot can happen in even a single microsecond: Earth orbits another 18.5 millimeters (about ¾ in.), light travels 300 meters (1,000 ft), proteins in our cells stretch and fold into complex three-dimensional shapes in order to carry out their life-enabling tasks.

What, then, of a nanosecond—a billionth of a second? Surely nothing could be that fast. And yet a microprocessor inside a desktop computer takes just a few nanoseconds to carry out an instruction, such as adding two numbers. In the time it takes you to blink your eye, a typical computer can do 900 million calculations. Suddenly Nicholas Fattu's experience starts to make sense: There are as many nanoseconds in a single second as there are seconds in 30 years—an entire lifetime of math performed by a human could be reproduced in a matter of seconds on your smartphone.

However, at the smallest measurable sizes, nanoseconds can seem like an eternity. Remember that when you see a red light, it's electromagnetically vibrating 400,000 times each nanosecond. The fastest switching transistors—the hardware that lies at the heart of a computer—operate in the realm of trillionths of a second, or picoseconds. Imagine living at this scale: Photons of light cover only one millimeter in a picosecond.

It takes the molecules in your eye only about a fifth of one picosecond—200 femtoseconds—to react to visible light. That light itself is constructed of wavelengths flip-flopping between electric and magnetic fields every two to four femtoseconds. These atomic speeds seem untouchable by humans, and yet today's fastest computers, which are measured in petaflops (quadrillions of operations each second), can complete a calculation in less than one femtosecond. At this rate, it's hard to tell a split second from a split-split-split second, but consider that there are as many femtoseconds in 1 second as there are seconds in 31.7 million years. Or, as *Scientific American* magazine put it, "More femtoseconds elapse in each second than there have been hours since the big bang."

In the late nineteenth century, Eadweard Muybridge stopped a racehorse in midrun by capturing a brief moment in time on a photographic plate, proving that all four legs were in the air at the same time. By the 1980s, scientists used the same techniques, but

The individual H_2O molecules in water have a slight attraction to each other, bonding for just a few picoseconds at a time. It's like a disco dance floor of constant movement, where molecules partner off, then break up and dance with others—the attractions keep it tight enough to form a mob but loose enough so that foreign molecules can easily wander through the liquid.

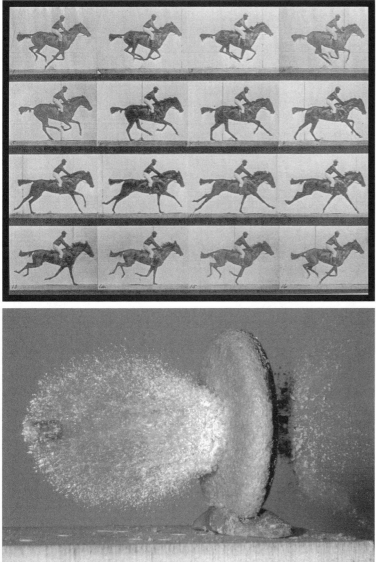

Eadweard Muybridge / Library of Congress

Jeff Krug

"For what is time? Who can easily and briefly explain it? Who even in thought can comprehend it, even to the pronouncing of a word concerning it? But what in speaking do we refer to more familiarly and knowingly than time? And certainly we understand when we speak of it; we understand also when we hear it spoken of by another. What, then, is time? If no one ask of me, I know; if I wish to explain to him who asks, I know not . . . How then can these two kinds of time, the past and the future, be, when the past no longer is, and the future yet does not be."

—Saint Augustine of Hippo, *Confessions*

▲ Eadweard Muybridge's camera stopped a horse in 1881. Today's cameras are faster than a speeding bullet.

this time freezing never-before-seen molecular reactions by firing femtosecond laser pulses like a camera strobe light—flashes of light one millionth of a billionth of a second. It's an extraordinary accomplishment, but individual atoms still appear as nothing but a blur at this speed; an electron has completed its entire virtual orbit around an atom in that single femtosecond. Clearly, if we're going to travel into the impossibly small world inside the atom, we need to look at even smaller slices of time, at the very limits of light, size, and even matter itself.

Quantum Time As far as we can tell, everything in our universe is tied to the speed at which light can travel, and light travels really, really, really fast: 299,792,458 m/sec; or 186,282 miles, 698 yards, 2 feet, and just over 5 inches. Everyone agrees it's far easier simply to memorize "about 300,000 km/s."

Remember that light may travel far slower than this, depending on what it's traveling through: in water, it's about 75 percent as fast; through a diamond, it's only 40 percent. Scientists can actually force charged particles to travel faster than light through a dense medium, which creates a bizarre electromagnetic shock wave called Cherenkov radiation. But even then, the particles can't go faster than the maximum limit of light in a vacuum. Of course, that limit refers to travel only through the four-dimensional medium called space-time, and it's possible that light (or particles, such as neutrinos) may be able to travel faster if we someday discover additional dimensions.

Nevertheless, our understanding of the speed of light, along with our current grasp of the size of the universe, leads to some interesting realizations. Foremost, you can forget those fantastic images from *Star Trek* or *Star Wars* where the stars suddenly race past as you move beyond light speed. Even if you could travel "Warp 7" (which geeks will immediately recognize as 656 times the speed of light), it would still take more than 2 days to get to the nearest star beyond our sun, and 152 years to get across our galaxy.

> "It's hard to imagine a more mind-stretching experience than learning, as we have over the last century, that the reality we experience is but a glimmer of the reality that is."
>
> —Brian Greene, physicist

The speed of light also causes some fascinatingly weird challenges, such as: If you're driving your 1972 Camaro at 99 percent of the speed of light and you flip on the headlights, what happens? The answer, in short, is that you see the light travel away from you at—you guessed it—the speed of light. This conundrum is due in part to the fact that the faster you travel, the slower you become—that is, time slows down for you relative to everything else. Plus, as you speed up, you actually get squeezed in length—you become shorter along the direction you're traveling, called the Lorentz contraction. These are infinitesimal changes at any speed we can actually imagine: Even if you were traveling at 42 million meters per second (95 million mph, or about a seventh of the speed of light), your length would contract by only 1 percent. But accelerate to just below the speed of light, and you would appear from the outside as though you were squished down to a speck's width—and while time would progress normally from your perspective, it would appear as though it had virtually stopped for you from the outside.

Again, these bizarre effects are meaningless at our human scale, but they absolutely must be taken into account in the realm of the atom, where particles and energy fields move and change unbelievably fast. In half a femtosecond—or 500 attoseconds—light travels 150 nanometers, about the size of a virus. A single attosecond is a billionth of a billionth of a second, so there are almost twice as many attoseconds in each second than there have been seconds since the big bang. But it still takes a full attosecond for light to traverse the length of three hydrogen atoms.

Some scientists have proposed that the shortest possible duration—the smallest slice of time that makes any sense or that we would ever need consider—should be called the chronon and measured as the time it takes for light to travel the diameter of a single atomic proton: about 6 yoctoseconds, or six millionths of an attosecond. Proponents argue that every event from the nuclear to the cosmic can be broken down into a series of chronon-long

The special theory of relativity states that an object would require an infinite amount of energy in order to accelerate to the speed of light. But it doesn't actually preclude things that are already faster than the speed of light. Some theorists believe subatomic particles called tachyons may exist that can travel *only* at superluminal speeds, not slower.

segments. It's a compelling idea, but unfortunately quantum physics has opened a world far smaller than the proton. How can we explain the strange workings of quarks and bosons if we're limited by chronons?

Which brings us to the Planck, the smallest possible measurement—anything smaller and our physics truly dissolves in a foam of random probability. The Planck length is a hundred billion billion times smaller than a proton: roughly 1.616×10^{-35} meters. A photon traveling at the speed of light would take one unit of Planck time to cross one Planck length: about 5×10^{-44} seconds—truly the ultimate "quantum of time."

The Trouble with Time Now that we have constructed a carefully calibrated measuring stick for time, with clear marks to help us understand and name any duration or speed, let's apply it to a simple question: How long is the current moment? Perhaps a moment is experienced at our human scale, near a tenth of a second? Or are moments chronon or Planck sized? Or perhaps "now" is longer than we think, lasting an eon?

The trouble, you see, is that we can discuss time all day, but no one—from the scientist to the spiritualist—really has any idea what he's talking about. Every inquiry into time ultimately distills down to a series of questions, like dregs at the bottom of a bottle. For example, here's another one: Is this minute as long as one we had yesterday? While you consider that, remember, too, that our only tool to measure time unfortunately uses circular reasoning: "Time passes at exactly one second per second." Worse, time has a way of slipping by in such a way that we cannot hold one duration up to another to compare them, like we could with two objects. And worst of all, while common sense tells us that time marches forward at the same rate everywhere, physicists now insist that's almost certainly not true. Instead, due to the force of gravity, unless you're lying down, your head is literally aging faster than your toes.

You can blame Einstein for pointing out this odd disparity. In classical physics, Isaac Newton's commonsense science of apples dropping and pendulums swinging, time is an absolute, fundamental structure, like a frame on which we can drape tautly drawn sheets of space. It's a comforting view of time, and one that usually works exceedingly well. But Einstein's theory of relativity points out that how time appears to us from our human perspective is not at all how time behaves everywhere. It's like someone saying, "One plus one equals two in most cases, but sometimes it's a bit more or a bit less."

Relativity tells us that time, intricately interwoven with space, is a flexible medium and that it stretches based on gravity and speed. This implies that two objects moving at two different speeds would travel through time at two different rates, which seems crazy . . . but it turns out to be true.

When the U.S government shot the first global positioning satellite (GPS) into space in the 1970s, no one knew for sure if Einstein's theories were right, but millions of dollars were at stake: If the onboard clock became inaccurate by more than about 25 nanoseconds, it would be effectively useless. However, relativity predicts that a clock traveling through space at the speed of an orbiting satellite should tick far slower than one on the ground—about 7 microseconds per day slower! That alone would cause enough of a problem, but Einstein also predicted that a clock at that altitude—farther from Earth's space-time-warping gravity—would speed up by 45 microseconds each day. The two relativistic effects combined would cause an error of 38,000 nanoseconds each day, enough that a GPS unit on the ground would provide an incorrect reading after just two minutes.

Fortunately, like good investors, the scientists hedged their bets, including a switch on the satellite that could be remotely thrown to enable or disable corrections for relativity. It didn't take long to realize that relativity worked exactly as predicted decades earlier: Speed and gravity affect time.

A clock at the top of Mt. Everest will pull ahead of one at sea level by about 30 microseconds per year.

In today's world of hyperaccurate measurements, scientists have found the same effects here on Earth. They can detect the slowing of time you experience when you ride a bicycle, or how you age faster when you climb a ladder. From our human scale, you never notice the effects, which add up to only billionths of a second over an entire lifetime. But the adjustments are significant when dealing with the ultrafast: Physicists working with atomic clocks must now correct for relativity even when they compare clocks on different floors of the same building. It's not that one clock is more accurate than the other; each clock, just like each of us, literally steps through time differently.

With that in mind, our understanding of "now" takes on an even stranger twist: Just as "here" means "where I am," "now" is intimately tied to "when I am"—time is always personal.

Consider, too, that our dependence on light leads to the realization that there is always a time delay between when something happens and when we can learn about it. If the sun exploded right now, we wouldn't—literally couldn't, due to the speed of light—know for about eight minutes. In other words, relativity says one object cannot affect another object without the passage of time. Add to this inevitable delay the fact that two events that may appear to happen at the same time from one point of reference will appear to be happening at different times from someplace else, and the whole idea of us ever truly grasping a "now" basically falls apart.

As gravity increases, time inevitably slows. Therefore, inside the infinite gravitational field of a black hole, from which even light cannot escape, time stops.

There's No Time Like the Present The dissonance and anxiety every sane person feels when confronted with these facts about time are normal; after all, time is fundamental to everything we do, everything we feel we are. And each of us has a choice: We can simply accept our intuitive sense of time and enjoy what little we have left, or we can stew, dissatisfied with the easy answers, longing for a clearer picture.

On the one hand, Eastern philosophy directs our attention to the present moment—that infinitely small and fleeting slice—as all we have. This is wonderfully encapsulated in the Hallmarkian poem "Yesterday is history, tomorrow is a mystery, but today is a gift—that's why they call it the Present." On the other hand, perhaps the scientist's credo was best summed up by the comedian George Carlin: "There's no present. There's only the immediate future and the recent past."

The truth, if we may use such a bold word, is almost certainly weirder, delving far into a world more commonly explored by philosophers and daydreamers.

Case in point: Physicists have thrown yet another wrench into the time machine. Whereas classical physics describes an absolute clock, and relativity proves there isn't one, quantum physics appears to point, once again, to some larger, external stopwatch synchronizing events throughout the universe. Of course, like everything involving quantum physics, its argument seems to defy logic: A measurement performed on one subatomic particle (such as a photon or an electron) appears to be able to affect another particle elsewhere, simultaneously—as if the second one knew what was happening to the first, ignoring apparently inconsequential things like space or the speed limit.

But what really bothers physicists is that when you lay out all the equations that appear to describe our universe, none of them specifies a "now," or even that there is a future distinct from the past. For a physicist, time does not pass, or flow, or fly—it just is: Past, present, future are all one thing, like a finished "timescape" on a canvas.

In this model—called eternalism—our limited consciousness keeps us from seeing the bigger picture, constraining us to a single moment in time. We hold these moments as special, because they're all we have, all we're able to sense. But eternalism also implies that we're locked to a predetermined future, like actors in a movie playing our parts toward the inevitable (but as yet unknown) conclusion. If this were true, and we were to believe it, would our human sense of urgency and curiosity, our drive, disappear, knowing

A month before his death, Albert Einstein wrote of the recent death of his longtime friend Michele Besso: "Now he has departed from this strange world a little ahead of me. That means nothing. People like us, who believe in physics, know that the distinction between past, present, and future is only a stubbornly persistent illusion."

that none of it really mattered after all? And if so, would that reaction simply have been predetermined, too?

Even a hardened scientist finds this complete lack of free will distasteful. So try this on for size: What if there were multiple universes—in fact a near-infinite number of universes being created every moment—allowing for every possible future outcome to every choice you make. In this model, called the Many-Worlds Interpretation, all time—every event from the Big Bang to the End of It All—remains fixed within each universe like a solid block. And instead of us moving forward through time in one universe, we each move seamlessly from universe to universe, without knowing it, providing an illusion that we're stepping through time. While this no doubt sounds like a bad plot of some science-fiction movie, it happens to be the leading argument among the top minds in academia.

However, one of the hardest problems to solve in any of these eternalist models is why we remember the past but not the future—that is, why we move forward along the arrow of time. If time is just another dimension, like length or height, then shouldn't we be able to traverse it any which way we please?

You may recall from school that the second law of thermodynamics insists that, as time progresses, things get more and more chaotic (entropy increases). So if you spill your milk, it will likely cause a mess. That does seem to imply that events always proceed from past to future. But bizarrely, that's the case only from our macroscopic perspective. If you watched the scene from the atomic level, every interaction along the way, every atom or molecule that jiggles or wags as the milk spreads, could be played forward or backward. After all, it's just as likely that a single molecule will move up or down, left or right, which means that it is technically possible that the milk could flow back into the glass, effectively stepping backward in time. It's only when you look at the mass of the liquid as a whole—the average tendency of trillions of atoms—that flowing "backward" becomes very, very improbable. It's so improbable,

> "He asked me if I knew what time it was. I said, 'Yes, but not right now.' "
> —Steven Wright, comedian

Reputable scientists believe that time travel is theoretically possible, by using wormholes in space, infinitely long rotating cylinders, and other clever tricks. However, to generate these kinds of conditions at anything larger than the subatomic level, we would likely have to generate as much energy as an exploding star.

in fact, that we gain this overwhelming sense of time inexorably marching forward.

Nevertheless, at the smallest scales, our conventional sense of time starts to break down completely. Even the concept of the absolute, unchangeable past ("What's done is done!") is beginning to fray. It's becoming clear that, at the quantum level, decisions made today appear to affect events in the past. As the physicists Stephen Hawking and Leonard Mlodinow point out, "The (unobserved) past, like the future, is indefinite and exists only as a spectrum of possibilities . . . The universe doesn't have just a single history, but every possible history, each with its own probability; and our observations of its current state affect its past and determine the different histories of the universe."

With this in mind, it's entirely possible that the future is affecting the past and present all the time without our knowing it, as subtle changes at the quantum scale aggregate to influence our macrocosmic world. For example, in 2009, when the Large Hadron Collider in Switzerland was beginning its groundbreaking work to uncover the hypothetical subatomic Higgs boson, it experienced a major failure and had to be shut down. At the time, "a pair of otherwise distinguished physicists," as the *New York Times* called them, suggested that this may have been caused by the Higgs boson itself, traveling back in time to stop the collider before it uncovered it, "like a time traveler who goes back in time to kill his grandfather." While this may seem absurd, it could explain why everything at the quantum level is necessarily blurry to us; the future needs plenty of wiggle room in order to pull off changes in the present without getting caught.

Ultimately, it's likely that everything we think we know of time— our memories, our carefully constructed calendars, our most precise measurements—is an illusion, a mirage generated like a hologram, to enable us to make sense of a universe stranger than anything dreamt of in any of these philosophies. As Einstein wrote, "The only reason for time is so that everything doesn't happen at once." It may

> "Whether I come to my own to-day or in ten thousand or ten million years, I can cheerfully take it now, or with equal cheerfulness, I can wait."
>
> —Walt Whitman, *Song of Myself*

Duration*

5.4×10^{-44} s	Planck time (shortest possible time)
0.3 ys	Lifetime of W and Z bosons
6 ys	Light travels diameter of atomic proton (1 chronon)
1 as (1 million ys)	Light travels length of 3 hydrogen atoms
12 as	Shortest laboratory laser pulse on record
320 as	Electron transfers from one atom to another in an electrical reaction
1.3 fs	One cycle of electromagnetic light between visible and ultraviolet light
200 fs	Fast chemical reactions (such as eye reacting to light)
1 ps (1 million as)	Half-life of a bottom quark
3.3 ps	Light travels 1 millimeter
1 ns	One machine cycle on 1 Ghz computer chip
2.5 ns	One million wavelengths of red light
5.4 µs	Light travels 1 mile in vacuum
22.7 µs	Length of a sound sample on an audio CD
5 ms	Bee wing beats once
8 ms	Camera shutter speed at 1/125 s
33.3 ms	One frame in digital movie
41.7 ms	One frame in a film movie
200 ms	Average human reflexes
30 cs	Blink of an eye
43 cs	Fastball travels from pitcher's hand to home plate
1 second	Human heartbeat; light travels 300,000 km
9.58 s	World record 100 m dash
10.5 minutes	About the half-life of a neutron outside an atom
1,039 seconds (17 minutes, 19 seconds)	Record time holding breath underwater
28.8 ks (8 hours)	Average human daily sleep requirement
86.4 ks	One day
29.5306 days (2.55 Ms)	Lunar month
40 days	About the longest a person can survive without food
125 days	Life of red blood corpuscle
23 Ms (38 weeks)	Length of human pregnancy
356.2422 days	Average solar ("tropical") year
27.7 years	Half-life of strontium 90
75 years (2.3 Gs)	Typical life span for a human being
90 years	Life span of anemone (longest-living invertebrate)
3.16 Gs	One century

122 years, 164 days	**Oldest human: French woman Jeanne Calment (1875–1997); that's 3.86 billion seconds!**
150 years	**Life span of tortoise**
164.8 years	**Neptune's solar orbit**
248.09 years	**Orbit of Pluto around sun**
550 years	**Time since humans learned to roast coffee**
31.55 Gs	**One millennium**
2,540 years	**Time since Buddhism founded**
6,000 years	**Time since humans learned to brew beer**
11,800 years	**Time since last ice age (Holocene epoch)**
25,784 years	**Earth's axis returns to same location (precession of the equinoxes)**
1 Ts (10^{12} s)	**31,689 years**
100,000 years	**Time since Homo sapiens appeared**
65 million years	**Cenozoic period (time since the dinosaurs died off)**
225 million years	**One rotation of the Milky Way**
710 million years	**Half-life of uranium 235 (relatively rare)**
1.26 billion years	**Half-life of potassium 40 (we have billions of these in our bodies)**
2.4 billion years	**Time since the Great Oxygen Catastrophe**
4.5 billion years	**Age of Earth**
4.51 billion years	**Half-life of uranium 238**
4.63 billion years	**Age of sun**
13.75 billion years (about 434 Ps)	**Age of universe**
10 trillion years (312×10^{18} seconds)	**Estimated life span of a red dwarf star**
100 trillion years	**Stelliferous era (time until all stars burn out or collapse into black holes)**
311.04 trillion years (about 9.8 Zs)	**Lifetime of Brahma in Hindu mythology**
10^{34} years	**Estimated half-life of proton**
2×10^{67} years	**Life span of a small black hole (mass of the sun)**
10^{100} years	**Time until stars, galaxies, black holes, and virtually all matter in the universe cease to exist**

*You can refer to the table of prefixes and their abbreviations on the last page of this book.

even be that the mind itself creates and manages the flow of time. We don't know, and may not even be able to know, the answer.

From Here to Eternity We live at a rate that allows us to experience the imperfect metronome of daily tides, of seasons changing, of trees growing. But we cannot watch the blossoming of a flower or follow a bullet's trajectory with our own eyes—those speeds are outside our spectrum of ability. Nevertheless, the worlds of the very fast and the very slow are just as real as ours. Does a turtle think it's walking slowly, or a hummingbird have any idea how quickly it's moving? Or does the hummingbird watch us in wonder at our snail's pace?

The neurologist Oliver Sacks has written movingly of a patient in a near-catatonic state who sat for hours in one position. Only much later did Sacks realize that the man was moving: He was wiping his nose, but at a rate that took an hour to bring his hand to his face. Another patient lived on the opposite end of the spectrum, having no trouble catching a fly in midair because it appeared to her that the insect was moving lazily. In neither case did the patient think their time was any different from ours. Today, doctors are starting to find that time—or, more specifically, a mismatch in our personal sense of time and the time moving around us—may be at the heart of a wide range of physical and psychological disorders, including Parkinson's disease, attention-deficit/hyperactivity disorder, and even some forms of autism and schizophrenia.

Time is, without a doubt, a crucial element of our consciousness and our ability to understand not just the world around us but also the extraordinary worlds we can track with instruments. After all, only through time can we experience change, and only through the aggregate of the tiniest and most subtle changes can the magisterial movement of galaxies be achieved.

Of course, time isn't reserved for just the mundane workings and measurements of atoms or eons. The twentieth-century Jewish theologian Abraham Joshua Heschel wrote that God was to be found not in sacred places but rather in sacred time. If we're searching for meaning, or for a greater connection to the universe in which we live, it's clear that we must move beyond a comfortable and passive acceptance of time's mysteries.

What if 13.7 billion years, or even a trillion years, is but a single lifetime, one of many thousands? Perhaps the age of our universe is the equivalent of a single Planck time in another, greater universe, and the cosmos we know, the billions of stars living and dying, is but a momentary blip into existence before it snuffs out, all in a moment too small to even measure. Does that mean our human scale, our human time, is any less important, any less glorious? It's all a matter of perspective.

EPILOGUE

Imagination is more important than knowledge.

–Albert Einstein

WE ARE THE ONLY SPECIES WE KNOW OF THAT CAN IMAGINE, SO imagine with me: Imagine being a neutrino that, while traveling through even the most dense rock, is in wonder at the vast spaces within and between atoms, like a spaceship traveling between the planets and the stars. Imagine a galaxy enjoying the delicious sensation of its several billion star parts twisting and turning, like an early morning stretch. Imagine a raucous midair party game, as one molecule passes on messages by playfully nudging its neighbors, propagating the sound wave that is about to tickle your eardrum.

Imagine you can watch two atoms as they exchange electrons in a slow dance of attraction, then turn to see the swirling flow of mountains rising and eroding through the swiftly shifting power of water and air across an eon of time. It's all true, it's all happening right now, though we can experience these visions only through our mind's eye.

And now, look around you. Like so many real-world adventures, one of the best parts of embarking on a journey into outer and inner space is the return home, where we can rest into our comfortable human-scale senses, enlightened by our new awareness that the world is far more complex, fun, and—yes—intimidating than we had thought. But now returned, how are we to understand our place in this vast spectrum?

We see a rose and attempt to understand its roseness by breaking it down—cell by cell, molecule by molecule, atom by atom—to its

> "Science cannot solve the ultimate mystery of nature. And it is because, in the last analysis, we ourselves are part of the mystery we are trying to solve."
>
> —Max Planck, physicist

essence, and we discover that it is miraculously constructed from star dust, remnants of supernovae billions of years ago. And not only is it the same stuff we're made of, we find that at a certain level it's difficult to determine where *it* stops and *we* begin.

But as we draw our vision back, enhanced by our newfound insights, we find an even more astonishing perspective: The elegance of a single rose is different from the beauty and magnificence of a rose garden of ten thousand bushes. Neither is more beautiful, neither is more real; both must be taken as part of the whole and yet wonderfully separate.

It's like exploring a single note played within a Bach fugue versus its place in the theme, versus the piece as a whole, versus the canon of Bach's works, versus the whole of Baroque music. Or, conversely, considering the attack of that single note, its duration, its tremolo, its resonance. Can the entire splendor of the Baroque period be captured in the resonance of that single note? Can the amazing clarity and richness of the single note be adequately reflected in the waves of millions of notes from that period? Or is each perspective, each range, as valuable, as rich, as sublime?

So it is in our lives, caught here in the middle world between large and small, cold and hot, slow and fast. We seem insignificant—less than a speck on a speck, and still hardly able to see, much less manage, the world of the atom or that of the universe. But appearances can deceive. Is a star any more important than we are because of its great size or power? A loud sound can demand our attention, but does it necessarily convey more than a soft one? We humans have something neither the cosmos nor the atom has: mind. We can appraise and appreciate; we are curious and cunning; we delve and dream. And who is to say that in the grand scheme of things—whatever collection of dimensions and multiverses we truly live in—these qualities of mind are not just as meaningful as mass or momentum?

You might even say that creativity and communication are our force carriers, the way photons act within and between atoms, or

> "I regard consciousness as fundamental. I regard matter as derivative from consciousness. We cannot get behind consciousness. Everything that we talk about, everything that we regard as existing, postulates consciousness."
>
> —Max Planck, physicist

gravity plays between the stars. It is only from our reference point in the middle that we are able to survey and consider a panorama of spectrums.

Unfortunately, it's also true that our minds may not be ready to grasp many of the clues we're uncovering. In fact, most of what we have learned about the universe is literally inconceivable—our brains do the math, we can logically accept and argue it, but it's too weird, too astonishing to truly grok. So on the one hand, there is no doubt that we know far more than people a century or two ago; but on the other hand, we find ourselves still somewhere in the middle of a spectrum of understanding, or even of sentience.

We so want to believe that if we understand the parts, we will understand the whole—and if we understand the whole, we'll finally understand our place within it. If only we could gather enough data, we'd be able to forecast the weather, or understand time's arrow, or explain the delicious laughter of a nine-month-old playing peekaboo. It's a powerfully compelling concept, but twenty-first-century science is waking up to the idea that it's not going to happen anytime soon—or perhaps ever.

We're at the tipping point where much of what we think of as "truth" is being overturned, and our children are growing up in a world of "what-ifs" and "we cannot knows." To manage, we must learn to be curious adventurers, letting our left brain crunch the numbers while developing our right brain's ability to poetically intuit these extraordinary realms, if not to find answers then at least to find meaning.

If we've learned anything, it's that beyond every horizon is another mountain, inviting us to learn more and delighting us with new vistas. We must simultaneously surrender into the mystery, celebrate our achievements, and strive for more. For that's what we humans do as we compare ourselves to others, making spectrums of our experience that inspire with an almost spiritual sense of wonder.

> "Penetrating so many secrets, we cease to believe in the unknowable. But there it sits nevertheless, calmly licking its chops."
> —H. L. Mencken, satirist

ACKNOWLEDGMENTS

"Even though you can't see or hear them at all, a person's a person, no matter how small."

—Dr. Seuss, *Horton Hears a Who!*

I AM DEEPLY GRATEFUL TO A WIDE SPECTRUM OF INDIVIDUALS and organizations for their inspiration, education, and cooperation. First of all, my thanks to my father who art in Austin, Adam Blatner, whose brainstorming and encouragement were invaluable. My agent, Reid Boates, who talked me into what became the hardest project of my career, and publisher, George Gibson, who believed I could pull it off. And to my editor, Lea Beresford, and designer (and longtime friend), Scott Citron, who were crucial in the process. Thanks, too, to Nancy Chamberlain, Cynthia Merman, and Nicole Lanctot for checking my work and cleaning this up, and to Lisa Silverman for shepherding the process beautifully.

Many thanks to the American Museum of Natural History, the King County Library System, and Amazon.com, who fed my habit. A shout out of thanks to the Lyons' Den, Peets Coffee, and Café Ladro, without whose "inspiration juice" this book would have had far fewer adjectives; to Daft Punk, Delerium, Garmana, and Jean Michel Jarre, who gave it a beat; to the makers of DEVONthink Pro, who helped me keep all the pieces together; and to the beautiful Inn at Langley, where the final words were typed. To my friends and family, including Gabriel, Daniel, Mom, Richard, Allee, Don, Snookie, Suzanne, Damian, Lucia, Alisa, Paul, Camille, Zoe, Edna, Ted, Ruth, Glenn, Jeff, Mark, and Anne-Marie.

And my deepest appreciation to my wife and partner, Debbie Carlson, who reminds me that words are important and life is magical.

ABOUT THE AUTHOR

DAVID BLATNER IS KNOWN WORLDWIDE FOR HIS AWARD- winning books, including *The Joy of Pi* and *The Flying Book*, and his lectures on electronic publishing. More than a half-million copies of his books are in print in fourteen languages. He and his wife and two sons live outside Seattle, Washington.

FOR FURTHER READING

For additional links, facts, and information, visit: www.spectrums.com

Selected Books

Asimov, Isaac. *The Measure of the Universe: Our Foremost Science Writer Looks at the World Large and Small.* New York: Harper and Row, 1983.

Carroll, Sean. *From Eternity to Here: The Quest for the Ultimate Theory of Time.* New York: Dutton, 2010.

Davies, Paul. *The Goldilocks Enigma: Why Is the Universe Just Right for Life?* New York: Mariner Books, 2008.

Greene, Brian. *The Fabric of the Cosmos: Space, Time, and the Texture of Reality.* New York: Vintage, 2005.

Joseph, Christopher. *A Measure of Everything: An Illustrated Guide to the Science of Measurement.* Ontario: Firefly Books, 2006.

Kaku, Michio. *Hyperspace: A Scientific Odyssey Through Parallel Universes, Time Warps, and the 10th Dimension.* New York: Anchor, 1995.

Potter, Christopher. *You Are Here: A Portable History of the Universe.* New York: Harper Perennial, 2010.

Robinson, Andrew. *The Story of Measurement.* New York: Thames & Hudson, 2007.

Streever, Bill. Cold: *Adventures in the World's Frozen Places.* New York: Back Bay, 2010.

INDEX

absolute hot, 132
absolute zero, 109, 111, 120
acre, 31
Adams, Douglas, 2, 7, 41, 154
Agni, 133
air pressure. *See* pressure
aleph null, 24–25
ampere, 139
amplitude, 79, 90
amplitude modulation (AM), 77, 79
AM radio, 77, 79
ana/kata, 58
Anaxagoras, 42
Andromeda galaxy, 42, 147, 151
angstrom (Å), 48
animals, sound and, 99, 100–102
Anthropocene era, 148
Aristotle, 84
Aronofsky, Darren, 22
Asimov, Isaac, 59
astronomical unit (AU), 21, 35
asymptotic limit, 120
atmosphere, 33, 69–70, 85, 96, 148–49
atom bombs, 129
atoms
  heat and, 107, 112, 120, 122–26, 128
  mass and, 51
  numbers of, 16, 20, 21–22, 23, 29
  size of, 48–49
  space and, 56–57
  speed of, 158
  supercooling, 120–25
atomic clock, 139, 162
atto-, 19, 184
AU (astronomical unit), 21, 35
Augustine of Hippo, Saint, 157
Avogadro's constant, 20
avoirdupois, 31
Azad, Kalid, 14

Bach, J.S., 172
background radiation in space, 118
bacteria, 29, 148
base-60 counting system, 137

Bateson, Gregory, 7
bats and echolocation, 100–101
BEC (Bose-Einstein condensate), 124–25
Beijing time, 141
Besso, Michele, 163
Betelgeuse, 40–41
Bhagavad Gita, 133
Bible, 29, 129, 138
big bang, 103, 118, 151, 159, 164
billion, 17
bioluminescence, 64
Birdseye, Clarence, 115
black-body radiation, 127
black holes
  gamma rays and, 71
  gravity and, 41, 42, 162
  Hawking radiation and, 152
  heat and, 132
  at quantum level, 57
  sound and, 102–103
  time and, 167
black lights, 64
blood cells, 46
Bloom, Howard, 1
Blue Marble (photo of Earth), 37
Bolt, Usain "Lightning," 144
Boomerang Nebula, 119
Borges, Jorge Luis, 149
Bose, Satyendra Nath, 123–25
Bose-Einstein condensate (BEC), 124–25
Bosenova, 124
bosons, 51, 165–66
Bradbury, Ray, 126
brain freeze, 118
Brookhaven National Laboratory, 131
Browning, Elizabeth Barrett, 66
BTU (British Thermal Units), 111
buckyball molecule
  (buckminsterfullerine), 48, 122
Burj Khalifa skyscraper, 32, 34

Cage, John, 88
Calabi-Yau manifolds, 58
calendars, 136–38, 140
calories, 111, 112
Camus, Albert, 97
Canis Major dwarf galaxy, 42

carbon dioxide, frozen, 115–16
Carlin, George, 5, 143, 163
Cassini spacecraft, 36
cells, 16, 18, 20, 29, 40, 45–49, 75
Celsius scale and Anders Celsius, 110,
    111–13
Cenozoic era, 148
Centauri star cluster, 40, 85
centi-, 184
centigrade scale. *See* Celsius scale
change and time, 133–36, 165
chanting, 104–105
chaos, 111–13, 164
char, 126
Cherenkov radiation, 158
chess, 14–15, 22
chronon, 159–60
chrononyms, 140
chunking, 20–21
climate change, 146
clocks, 138–39
color, 56, 74–76
color blindness, 76
comparisons, 1–2, 13
complex numbers, 27
condensation, 115
cone cells, 75–76
cookie and bullet, 157
cooling, 113–25
Coordinated Universal Time (UTC), 144
Cornell, Eric, 123–24
cosmic time, 150–52
counting and counting systems, 11–12,
    13, 30, 31–32, 138
cryobiology, 117
cubit, 30

Dalí, Salvadore, 68
Dangi, Chandra Bahadur, 32
dark energy, 55
Davis, Philip, 24–25
Dawkins, Richard, 2, 23
day, definition of, 138
daylight savings time, 142
dB (decibel), 92
dB drag racing, loudness of, 93
decaffeination process, 128

decibel (dB), 92
DeGeneres, Ellen, 72
Democritus, 127
diffraction of sound, 98, 100
diffusion, 146
dinosaur extinction, 148, 149, 150, 151
DNA, size of, 48
double-slit experiment, 73
Draper point and John Draper, 126–27
dry ice, 115–16
duration, 136, 159, 166–67
    See also time

e (number), 25, 26
E.coli, 29, 47
eardrum, 97
Earth
    distance from sun, 35, 38–39
    rotation of, 138, 147
    size of, 34, 36
    speed of, 150
    wobble of, 146–47
echoes, 98, 100
echolocation, 100–101

Edison, Thomas, 6
Egyptian calendar, 137
Einstein, Albert
    Bose-Einstein condensate and, 124–25
    electromagnetism and, 81
    on imagination, 171
    on numbers, 16
    relativity and, 54–55, 132, 161
    spukhafte Fernwirkung and, 57
    on time, 163, 165
electromagnetic radiation (EMR), 61–63,
    64, 69, 74, 85
    See also light
electromagnetism, 35, 48, 80–81
electrons
    clocks and, 139
    described, 50–51
    ionization and, 128
    light and, 63, 64, 69
    microscopy and, 47
    numbers of, 15
    speed of, 158

comprehending, 15–20
exponents, 12–15, 23
negative numbers, 23–25
nomenclature for, 13, 19
"outside the box," 25–26
prime, 22–23
very large numbers, 21–23

oldest organisms on Earth, 145
olympicene, 50
Oort cloud, 37, 39
Oppenheimer, Robert, 133
Orozco, Luis, 122
outer space, 34–45
  beyond our solar system, 40–45
  light-years in, 37–40
  size and distance in, 34–37
  temperature in, 118–19
  See also space and space-time
oxygen, liquid, 117
ozone layer, 69–70, 149

Pa (Pascal), 90, 92–94
parallax, 38
parsec, 38
Pascal, Blaise, 9
Pascal (Pa), 90–91, 93–94
pattern recognition, 11–12
peta-, 19, 184
petaflops, 15, 156
Phanerozoic geologic eon, 148
phase change, 112–13, 115–16, 123, 127
Phelps, Michael, 154
Philosophy of Space and Time, The
  (Reichenbach), 54–55
phosphors, 64
photons
  laser cooling and, 122
  light and, 72–75, 79, 81, 120, 127
  speed of, 156, 160
photosynthesis, 148
π, 25, 26
π (movie), 22
pico-, 19, 184
Planck, Max, 1, 132, 150, 171
Planck length, 52, 57, 58, 160
plasma, 128–29

plasma cutters, 128
Plato, 8, 103
Pluto, 36, 37
Poincaré dodecahedron, 58
pollen, size of, 46
Potter, Christopher, 40
pound (unit of measurement), 31
pressure
  heat and, 113, 114, 115, 127–28, 129
  light and, 71, 72
  sound and, 83, 89–92, 93–94, 97, 98, 102
  time and, 140
prime numbers, 21, 22–23
Principia Mathematica (Newton), 53
protons, 49, 51, 52, 131, 159
Pythagoras, 94, 98

quantum mechanics, 81, 123
quantum physics, 121, 125–26, 160, 163,
  165
quantum time, 158–59
quark-gluon plasma, 132
quarks, 51, 52, 131

R136a1, 71
radiation, 69–70, 79, 118, 127, 152, 158
  See also electromagnetic radiation (EMR)
radio waves, 67–68, 77
rainbows, 76
red dwarf stars, 40, 152
red light, 66, 67
redshift, 56
refrigeration, 114
Reichenbach, Hans, 54–55
relativity, 54, 132, 159–63
religion and sound, 104–105
resonance, 105
retina, 75–76
rhinovirus, 47, 49
rock concerts, 92–93
rocket fuel, 117–18
rod cells, 75–76

Sacks, Oliver, 168
Sagan, Carl, 24, 36, 67, 151
Sagittarius A, 41
Saint-Exupéry, Antoine de, 77

telluric currents, 67
temperature. *See* heat
tera-, 19, 184
tesseract, 58
tetrachromats, 76
tetration, 22
theory of relativity, 54, 161–62
thermodynamics, laws of, 113, 164
thermometers, 108–11
Thilorier, Charles, 115–16
Thomas, Dylan, 133
Thomson, William, 111
timbre, 4, 104
time, 135–69
  change and, 135–36, 168
  cosmic time, 150–52
  flow of, 162–65
  geologic time, 145–50
  gravity and, 160, 161
  human time, 142–45
  measurement of, 136–42
  perception of, 165–66
  quantum time, 158–60
  relativity, 161–62
  space and, 53–59
  speed and, 136, 144–45, 153, 160, 161
  very fast things, 152–58
time travel, 164
time zones, 140–42
transcendental numbers, 25
trees, oldest on earth, 145
triboluminescence, 64
trichromats, 76
trillion, 17
Tuoriniemi, Juha, 125
Turrell, James, 70
Twain, Mark, 29

ultra low frequency, 68
ultrasound, 95, 100–101
ultraviolet light, 64, 66–67, 69–70
universe, 37, 45, 80–81, 151
Uranus, distance from sun, 36
Ussher, James, 140
UTC (Coordinated Universal Time), 144

van Loon, Hendrik Willem, 24
Venus, 35, 72, 144
vibrating strings, 52
violet light, 67
Virgo supercluster, 42
viruses, size of, 47, 49
visible light spectrum, 62, 74–76, 80
Voyager 1, 35
VY Canis Majoris, 41

Wadlow, Robert, 32
Warlpiri (language), 12
war time, 142
warp speed, 158
Washington Monument, 32
water, measuring, 15, 20, 23
water molecules, 49, 156
wavelength, 64–68
  See also light; sound
wave-particle duality, 72–74
weather conditions and sound, 96
week, length of, 138
whales, singing of, 95, 101
Whitehead, Alfred North, 59
Whitman, Walt, 165
Wieman, Carl, 122, 123, 124
wonderment, 7–9, 168–69
Wordsworth, William, 66
wormholes, 56, 164
Wright, Steven, 107, 164

X-43A aircraft, 90
X-rays, 47, 64, 67, 69, 70–71, 77, 79–80

year, seconds in, 136
yocto-, 19, 184
yotta-, 19, 184

zetta-, 19, 184
Zulu time, 144

Prefix		Size	Name
yotta-	Y	10^{24}	septillion
zetta-	Z	10^{21}	sextillion
exa-	E	10^{18}	quintillion
peta-	P	10^{15}	quadrillion
tera-	T	10^{12}	trillion
giga-	G	10^{9}	billion
mega-	M	10^{6}	million
kilo-	k	10^{3}	thousand
hecto-	h	10^{2}	hundred
deca-	da	10^{1}	ten
deci-	d	10^{-1}	tenth
centi-	c	10^{-2}	hundredth
milli-	m	10^{-3}	thousandth
micro-	μ	10^{-6}	millionth
nano-	n	10^{-9}	billionth
pico-	p	10^{-12}	trillionth
femto-	f	10^{-15}	quadrillionth
atto-	a	10^{-18}	quintillionth
zepto-	z	10^{-21}	sextillionth)
yocto-	y	10^{-24}	septillionth